I0465244

Reincarnation, Maternal Impression, and Epigenesis

Reincarnation, Maternal Impression, and Epigenesis

Milton Brener

Copyright © 2016 by Milton Brener.

Library of Congress Control Number:		2016919106
ISBN:	Softcover	978-1-5245-5957-1
	eBook	978-1-5245-5956-4

All rights reserved. No part of this book may be reproduced or transmitted in any form or by any means, electronic or mechanical, including photocopying, recording, or by any information storage and retrieval system, without permission in writing from the copyright owner.

Any people depicted in stock imagery provided by Thinkstock are models, and such images are being used for illustrative purposes only. Certain stock imagery © Thinkstock.

Print information available on the last page.

Rev. date: 11/16/2016

To order additional copies of this book, contact:
Xlibris
1-888-795-4274
www.Xlibris.com
Orders@Xlibris.com
738551

CONTENTS

Explanation of Cover Illustration

The figure on the cover is a representation of a 'nucleosome,' a segment of DNA wound around eight 'histone- protein' cores, shown here in different colors. It is like thread wrapped around a spool. The thread here is a strand of DNA. The spool is the 8 histone-proteins, namely, two each of four major types, packed into the nucleus of the cell. Remarkably, that nucleus is only 10 micrometers in diameter. A micrometer is .0001 of a centimeter. That improbable packing is one of the tasks of the proteins. The strands of DNA running throughout our bodies in all of our 50 trillion cells would equal 300 round trips between Earth and the sun. The purple 'tail' shown on the upper left of the diagram is the beginning of one of the strands. Together they connect millions of nucleosomes. The role of the nucleosome as a general gene repressor has been convincingly demonstrated. They are thought by some scientist to carry the epigenetically inherited information in their core histones.

Annunziato, A. (2008) DNA Packaging: Nucleosomes and Chromatin. Nature Education 1(1):26

**To: Eileen, and
To Lisa, Ann, Neil and Mat**

Special thanks to my wife, Eileen, whose professional expertise caught many errors and awkward phrasing before submission to the publisher.

Special thanks also to my daughter Dr. Ann Brener, for her generous gift of treasured books. They stimulated me to think more about, and to write about, this subject.

Introduction

Skilled investigations by medical professionals and others have established some important and highly interesting facts. They have found that at least several thousand children in recent times, world-wide, have been born with mostly, sometimes remarkably, accurate memories of lives from the past.

In two volumes on this subject I centered on the possible role of entanglement, an aspect of quantum physics, as the mechanism by which this occurs. Entanglement refers to relationships between atoms, extraordinarily tiny objects, which are the ultimate building blocks for everyone and everything. The characteristics of entangled atoms complement each other and a change in one causes a corresponding change in the other, instantaneously, at once, no matter how far apart, even if light years, they may ultimately be. The relationship may last for eons, longer than the projected life of the solar system.

With regard to the children who are born with memories of another life, by many of the children themselves, and often by others, they have been considered to be reincarnations in the traditional sense, namely the rebirth of an individual who lived in earlier times, with all of that individual's memories, personal qualities, mannerisms, thought processes, and other traits that make each of us unique. All such traits are often subsumed under the concept of 'soul.'

I believe however that the available evidence is somewhat more limited. It shows that these children have in fact 'inherited' the memories and associated emotions of, usually, predeceased persons, whom the researchers in this area, beginning with Dr. Ian Stevenson, term the 'prior personality.' The persons who claim to remember other lives are termed 'subjects.'

The term 'inheritance' used by me, not by the researchers, does not limit the prior personality to a member of the subject's family. It seems, rather, at least most often, a matter of chance. The subject and the prior personality may not be related at all, and often may never have even known each other. The persons involved and even the investigators frequently refer to these occurrences as 'reincarnations,' a term I would not use for these cases, but I nonetheless do so here, primarily to avoid confusion.

Though I do not claim that I have proved entanglement to be the engine that drives this phenomenon, I do believe that it is, in the present state of knowledge, a possibility. That is, I believe the organs of memory, the atoms of which are probably mostly 'entangled' with each other and most of which will survive us by many billions of years, can become part of a new fetus or infant. Those memories would be among the first mental pictures that the mind of the fetus or infant encounters, and among the first utterances of the child as it begins to speak. Usually by age 7 or 8, these children become integrated into their current surroundings and friendships, and memories from another life are mostly, sooner or later, forgotten or suppressed.

There remains however at least one further matter, one that is the subject of this book, and that on first blush seems anomalous and ill-fitting to the entire idea of survival of purely mental images. The seemingly anomalous matter arises from the fact that so many of these subjects are born with physical indicia of wounds and injuries, birth marks or birth defects. They are significant as they often bear close similarity, sometimes remarkably so, to wounds or injuries from which the prior personality, when identified, was found to have died, or suffered during his or her lifetime.

Even in the 22 cases I have summarized in my previous two books on the subject of reincarnation, examples of this phenomenon abound. Of the twelve cases summarized in the earlier book, *Our Quantum world and Reincarnation*, six contain examples of it. I made only the barest mention of these birthmarks or defects in some of them, no mention at all in others. I felt those matters to be unnecessary for an understanding of the cases and perhaps a bit too much to expect most readers to digest together with the other strange and unfamiliar phenomena.

Further, I had not yet given sufficient thought or research with which to support my growing suspicion that epigenesis may have played a role. I have extracted here the pertinent facts that illustrate those aspects of the cases in my two prior volumes plus many other cases involving the same phenomenon. It may appear that I have overdone it with the number of cases cited. I rely upon that old saw that extraordinary claims require extraordinary proof. Beginning with chapter 7, I will explain my suggestion that these phenomena can possibly be driven by the science of epigenetics.

NB: By most standards, this book is short in length as is the bibliography. Hence instead of the usual endnotes, I have, in the body of the text, in relatively few places, referred to the source of my material, with the name of the book or document and page number instead of the customary mention of the author first. The bibliography is short enough that, for those who are interested, it would be a simple matter to identify the volume with all necessary information, including name of the author. I have cited most often Dr. Ian Stevenson's two volumes of *Reincarnation and Biology* and have taken the liberty of referring to those volumes as *R and B*. This was all meant to simplify, not complicate, a good intention that does sometimes backfire.

Figure 1: Dr. Ian Stevenson 1918 – 2007; chair of the Department of Psychiatry, University of Virginia, from 1957 to 1967.His reincarnation research began in 1960, and resulted in multiple authoritative volumes on the subject.

PART I

Evidence for Survival of
Some Physical Defects

Chapter 1

Physical phenomena from
previously Summarized Cases

The first example of the phenomena at issue here comes in the very first of the cases in my first book on reincarnation, namely that concerning William McConnell, a New York policeman. Trying to prevent a robbery, he was shot six times, the fatal bullet entering his back and penetrating his left lung, heart, and main pulmonary artery.

One of his daughters, Doreen, five years later bore a son she named **William**. From the age of three the child gave evidence of familiarity with details of the life of his grandfather. Even earlier, shortly after birth, It was found that the infant's valve of the pulmonary artery had not adequately formed, blocking blood from flowing through it to the right ventricle to the heart. This required him to take medication indefinitely. The location of the birth defects, according to Dr. Jim Tucker, about whom we will hear more shortly, was very close to his grandfather's corresponding wounds.

*　　*　　*

Purnima Ekanayake was born in Sri Lanka (*Quantum World*, p.20). From birth she had light colored birthmarks on the left side of her chest and lower ribs. Beginning just before her third birthday she began to speak of a previous life as a male incense maker in a town about 145 miles from her hometown. This is one of a number of cases involving cross-genders. By her 6th birthday she had made at least 20 statements concerning the 'former life' including the claim that she was selling incense sticks on a bicycle when she was killed in an accident with a big vehicle.

Investigation by her school teacher revealed that a man in that town named Jinadasa Perera was killed by a bus while taking sticks to market on a bicycle about two years before Purnima was born. Investigation by Dr. Erlendur Haraldsson revealed that 14 of 20 detailed statements she had made about Jinadasa and of her life in the other town were correct; 3 were incorrect, and accuracy of 3 could not be determined.

Dr. Haraldsson also examined the autopsy report of Jinadasa, which described fractured ribs on the left, a ruptured spleen, and abrasions diagonally across the chest to the left lower abdomen. The injuries coincided closely with the birthmarks on Purina's chest and ribs.

Figure 2: Erlendur Haraldsson: professor emeritus of psychology on the faculty of social science at the University of Iceland. In association with Dr. Stevenson and others he has researched and published on the subject of reincarnation in various journals of psychology and psychiatry

*　　*　　*

Selim Fesli was a farmer in the village of Hatun Köy, Turkey. While resting in a field after work one day he was fatally shot and wounded by one,

İsa Dirbekli, who was hunting rabbits. Dirbekli mistook Selim for a rabbit and shot him at close range. The pellets entered the right side of his head and he died a few days later. Several months thereafter **Semih Tutuşmuş** (*Quantum World*, p.29), the son of Ali and Karanfil Tutuşmuş was born in the neighboring village of Şarkonak, two kilometers from Hatun Köy.

The father, Ali, had known of the circumstances of the death of Fesli, and both of Semih's parents saw their child as the reincarnation of Fesli. At the age of about 18 months the child did in fact begin to speak as if he were Fesli, his first words being İsa Dirbekli, the killer of Fesli. The child's father, Karanfil, tried to discourage him and beat the boy severely to try to stop his repeated journeys to his remembered home.

There followed nonetheless numerous statements by the child, Semih, about Fesli, his life and his death. He correctly gave the name of Selim Fesli's' wife and all six of their children, and remembered being shot in the right ear by İsa Dirbekli.

Why did the parents suspect so quickly upon the birth of Semih that he was the reincarnation of Fesli? It was because the infant, on the right side of face, had only a narrow linear stub of skin, with no external ear. The left ear was normal.

Semih bore much bitterness and ill will toward İsa Dirbekli, and accused him of shooting him deliberately, but at the time of Stevenson's last meeting with Semih, in 1977, he had given up any intention of revenge. His ear had been restored by surgery to a normal looking one while in the army between ages 16 and 18. Stevenson writes that the birth defect of the ear has an incidence of one in between 20.000 and 30,000 births. He said further that there was an exact correspondence between Semih's malformed ear and the wounds in Salim's head

*   *   *

The next case may raise questions as to whether birthmark or mutilation evidence can be sufficient where it is the entire, or the most significant, evidence of reincarnation. There was, here, minimal evidence of memories from the prior personality, who was a half-brother named Kevin. The subject, **Patrick Christenson** (*Quantum World*, p. 33) was born in Michigan in 1991. Upon first glance, the mother felt there was a connection between him and her first son, Kevin, who had died twelve years earlier of cancer at age 2 in 1979.

Kevin had begun to limp when he was 18 months. Shortly thereafter he fell and broke his left leg. Medical workup included a biopsy of a nodule on his scalp above his right ear. He was diagnosed with metastatic cancer, and upon a bone scan many abnormal sites were found.

Due to a tumor, Kevin's left eye was protruding and bruised. Chemotherapy through a central line on the right side of his neck resulted in his neck becoming flushed and slightly swollen on several occasions, but he was eventually discharged and sent home. He continued on outpatient treatment but returned to the hospital five months later with a fever and blindness in his left eye. He was treated with antibiotics and discharged from the hospital. He died two days later at two years and three weeks.

Kevin's parents had separated before his death and the mother remarried twice. In her first new marriage she gave birth to a daughter and in the second to a son, Jason, before the birth of Patrick, our subject in this case. What made the mother think that Patrick was a reincarnation of Kevin? She had no such suspicion regarding the older son, Jason, likewise by the second of her remarriages, hence likewise a half-brother of Kevin. Patrick was born twelve years after Kevin died. On Patrick she immediately noticed three defects that matched those of Kevin when he died. Other evidence soon became apparent.

Twelve years after the death of Kevin, and immediately upon seeing Patrick, his mother noticed that he had a slanting birthmark with an apparent small cut on the right side of his neck, the same location where Kevin's central line had been used for chemotherapy. There was also a nodule on his scalp above the right ear, where Kevin's biopsied tumor had been, and a corneal glaucoma in the left eye which caused him, like Kevin, to have very little vision in that eye. When he began walking he limped, favoring his left leg.

Between the ages of two and three he often asked his parents for toys he claimed to have left in his "earlier life." He was just over four years old when he began saying things that his mother felt related to the deceased Kevin. He said he wanted to go back to their previous home, which he correctly described as orange and brown. He asked his mother if she remembered him having an operation, to which she replied that he did not have any surgery. He then pointed above his right ear, the same place where Kevin's biopsied nodule had been located. He said that he could not remember the surgery itself as he was asleep when it was performed.

Pictures of Kevin were not normally displayed in the home, but on one occasion when they were, Patrick said it was a picture of him and that he had been looking for it. On occasions Patrick spoke of episodes of which he could not have had firsthand knowledge, but accurately described episodes from the life of Kevin. They took Patrick to the home that Kevin had shared with his mother, but he did not make any statements clearly indicating recognition. When Patrick was five, Doctors Stevenson and Jim Tucker were called in to investigate.

In writing about this case, Dr. Tucker explained that in same-family cases, there is always the possibility that the child had heard others discussing the deceased. But the three birthmarks were not subject to that type of influence.

Figure 3: Dr. Jim B. Tucker: author and Medical Director of the Child and Family Psychiatry Clinic, University of Virginia. He has researched and published widely on the subject of reincarnation

Upon visiting the home, doctors Stevenson and Tucker photographed the birthmark on Patrick's neck. It was a 4-millimeter dark slanting line on the lower part on the right side of the neck, and gave the appearance of a healed cut. They also palpated, and thus confirmed the nodule on his head, which at this time was small and very hard to see. They could also see the opacity in the child's left eye, and obtained copies of his eye examinations. By watching him walk they confirmed the slight limp, though there was no medical condition to explain it. They obtained Kevin's medical records documenting the history, including the lesions that appeared to correspond to Patrick's birthmarks.

Is the correlation between the surgeries, surgical scars, glaucoma, limp of the presumed prior personality, the corresponding birthmarks, and reduced vision and limp on the subject sufficient to support a conclusion of reincarnation? To many laymen, as well as scholars, it may all seem like entirely too much to dismiss as coincidence.

*   *   *

There are several other cases demonstrating the same phenomenon from *Something Survives,* my second volume on this subject.

One such case involves a population of about 200,000 in Israel known as the Druse. Like many of the world's populations they are firm believers in reincarnation. In this case (p. 39) a three year old boy had on his upper forehead from birth a red mark stretching to the center of his head. The boy claimed that a man had killed him in a prior life. He said he did not remember the name of the murderer nor his own name in the earlier life. As per the custom in that population, an inquiry was conducted by 15 respected elders including the boy's father. The only non-Druse in the group was an eminent doctor, named Eli Lasch.

When the search party entered the third village the boy identified it as the one where he had lived. He then also suddenly remembered his own names and those of his murderer. One of the elders of that village claimed to have known the man the boy named, adding that he had disappeared four years previously.

As the group traversed the village, the boy identified his house. An interested crowd was starting to gather. The boy walked up to a man and called him by name. The man acknowledged that was he. Said the boy, "I used to be your neighbor. We had a fight and you killed me with an ax." According to Dr. Lasch the man turned white as a sheet. The boy added "I even know where he buried my body."

The group, followed by many interested bystanders, followed the boy into a nearby field. The identified man was asked to come with them. The boy stopped in front of a pile of stones and said that the man had buried him under them. He also pointed, to where he said the ax was buried. The stones were removed and the skeleton of an adult man with a slit in front of the skull was found, whereupon the identified man admitted to the crime. The ax was also found where the boy said it had been buried. The group did not hand the man over to the police but told Dr. Lasch they would decide his punishment. What it was we are not told.

*   *   *

In the early 1950s Sri Jageshwar Prasad, a barber in Kanauju, India, District of Chhipatti heard of a boy born in a nearby district in July 1951. This was six months after the murder of one, Ashok Kumar, age 6, who usually had been called Munna. The boy born in July 1951, named **Ravi Shankar,** described himself as having once been the son of Jageshwar, a barber in the District of Chhipatti. The facts surrounding the murder of Munna had been well known.

On January 19 1951, a boy, named Ashok Kumar, nicknamed Munna, age 6, had been enticed away from his play and murdered with a knife or similar instrument. The severed head of the boy and some of his clothing was identified by his father, Jageshwar. Ravi Shankar described himself as the son of Jageshwar, a barber in the District of Chhipatti. He named the murderers, the place of the crime and some details of the life and death of Munna.

In this case, there is ample evidence of reincarnation apart from the physical, beginning with the insistence of Ravi between the ages of two and three for the toys, accurately described in detail, which were in the house of his "earlier life." Relevant to our focus here however, it is germane to note that Ravi's mother saw that Ravi had a mark from birth, first noted by her when the child was three to four months old, which seemed similar to a knife wound across the neck. When the child began talking he claimed it was from his wound from the murder of himself as Munna.

*   *   *

Chapter 2

The History of an Idea:
Maternal Impression

Stevenson sets out in Volume 1 of his *Reincarnation and Biology* a detailed history and examination of the concept of maternal impression. The heading of that section itself tells us much: *The Possible Influence of a Pregnant Mother's Images on her Embryo and Fetus.* He further explains that "A shock or other strong impression on a pregnant mother can produce a mark or defect in her baby." It is, he continues, not only a visual image, but sometimes a vivid verbal description that may cause the same result on the fetus or infant.

Stevenson claimed that evidence developed in more recent times made the belief that mind could act upon matter more plausible than previously thought. In the 20th century there was however, an indisputable decrease in the number of reported cases of it, at least in the western world. He expressed the opinion that the decline of belief in maternal impression resulted from the "increasingly materialistic view of human nature." He gave other reasons, the most cogent one of which was that any effect maternal impressions may have on fetuses, "occurs only among mothers and fetuses who are especially susceptible —either at a psychological level or a physical one."

Whether the dramatic decrease in such reports in the 20th century is due in large part to an increasingly materialistic view of human nature as

Stevenson suggests, it must certainly be true that his other explanation, the degree of emotional response to the original stimulus, rates strong consideration (*R and B*, vol. 1, pp.104, 173). It is said since early times that these marks, deformities, and evidence of wounds were the result of 'maternal impression,' and were described by Dr. Stevenson as "the belief that a shock or other strong impression in a pregnant mother can produce a mark or other defect in her baby."

Stevenson, has also written that though maternal impression is less common now in western countries, it is still widely accepted in other parts of the world. He explained further that women now may be less frightened by seeing deformed or marked persons and that persons with such marks or defects now receive more effective treatments. It appears that the psychological aspect of the phenomenon can hardly be exaggerated, and the less fright there is from such experiences, the fewer such cases there will be; and the fewer cases there are, the less frightened the women will be of adverse effects on the unborn child.

Though maternal impression was the earliest term attributed to this belief, it is now the subject of a more scientific analysis and of evidence that the phenomenon is not limited to pregnancies or to relatives. Beginning in the 18th, and continuing well into the 19th Century, controversy over the efficacy of the idea of maternal impression raged within the medical profession. It was defended by learned physicians, but dismissed as 'crude superstitions' by others of equal stature.

Stevenson also gives us a history of the controversy, mainly within the medical profession, as to the validity or invalidity of the entire concept. It is not absolutely necessary to know this history to follow the debate. It may however throw some light and greater understanding of the controversy, and some may find it interesting and worth knowing for its own sake.

Physicians of the 18th and 19th Centuries who defended its validity hoped to find nerve connections between the uterus and placenta that would in some manner convey a mother's mental impression to her fetus. These conjectures proved to be without substance, as were equally baseless suggestions of transmission of blood between the mother and the fetus. (*R and B*, Vol I, p. 105).

Is there any reasonable, logical explanation that could account for identity of birthmarks of the subject with wounds or marks on the body of the prior personality within the framework of the explanation for survival of memories as summarized in my two previous books? The answer, based on research by Ian Stevenson, and later analysis by Dr. Jim Tucker, is in the affirmative. It is an answer arising from an often used, but woefully underestimated, *bon mot*, namely 'mind over matter.' It may be surprising to most, how extensive and widespread is the evidence of that simple phrase.

Stevenson tells us that according to writers in Ancient Greece and Rome, those cultures "assumed the reality of maternal impressions." It was a belief that continued unabated at least until questioned in the 17th Century when a writer, Gervase Markham, scoffed at the belief (*R and B*, vol. 1 p. 104). In the 18th Century, the belief was defended by some scholars, but dismissed by others as nonsense.

One, J. Elliotson, in the mid-19th Century, is quoted by Stevenson as stating that "All my medical teachers dismissed the ideas with contempt." The German physiologist, Johannes Müller, gave his reasons for rejecting the belief, including lack of appropriate physical connection between mother and fetus, and the large number of pregnant women who were frightened, and expected babies that were deformed, but upon birth were quite normal. He dismissed the idea derisively as akin to animal magnetism.

His rejection did not extinguish the interest of many other physicians. The belief continued to find support from scientists of competence and eminence. Throughout the 19th Century there were several hundred reports of supporting cases published in medical journals and books. (p. 106). They were invariably answered by other authors who often raised the usual objections described by Müller. Perhaps one of the most forward looking defenses of the concept was penned by one A. Meadows in 1865 (*R and B*, vol. I, p 1162,).

Acknowledging the absence of any neural connections between the mother and her fetus, he suggested that the cases and the data itself "force upon us the conviction that the mind does in some mysterious way operate across matter." In 1870 one G.J. Fisher published a response based upon his own research enquiring of mothers' fears before delivery concerning deformity in their offspring. By far the larger number of mothers expressed such fears, yet only three among the babies showed any abnormality.

F. Barker, and S.C. Busey, both in 1887 wrote a rejoinder to Fisher. The former cited numerous examples of similarities between an unusual stimulus to the mother and a subsequent birthmark or defect on the child. Busey reviewed Barkers' paper and added 41 more cases, then answered the inevitable issue of chance: "The element of chance is eliminated by the great variety of causes with corresponding effects; that is, the circumstance producing the impression is different; yet in each case the effect is, to a greater or lesser degree, in correspondence with the causal circumstance."

A lengthy and thorough review of the issue was published by W.C. Dabney in 1890. His conclusion: "Maternal impressions account for few of all cases of birth defects. . . They are *one* of the causes of defects or deformities, but by no means the only cause." Stevenson noted that Dabney was undaunted by inability to find the process that could mediate between the mental impression and the related birthmark or birth defect." (*R and By*, vol.1, p. 107).

This did not silence the skeptics. They often protested even the publication of such reports and countered with cases of women who had been frightened, and were expecting malformed offspring but gave birth to normal infants without any such marks or deformities. J.W. Ballantyne in 1890 pointed to circumstances that "compelled the belief" that there was more than the elements of chance or coincidence involved. But this was becoming a losing battle.

Throughout the 20th Century, reports of various theories justifying belief in maternal impressions declined dramatically, often with derision. In the same year, 1890, Dabney had written that "thinking men came to doubt the truth of those things which they could not understand." He also quoted a thought by Pierre de Laplace 75 years earlier that "We are so far from knowing all the forces of nature and their processes that it would show little wisdom to deny phenomena just because we cannot explain them in the present state of our knowledge." (*R and B*, vol.1,p.108).

Chapter 3

Evidence Supporting
Maternal Impression

Stevenson first thought that maternal impression was the only means by which such a result could occur. By the time he completed his major work, twenty years in the making, he seems however to have realized that the phenomenon was much more extensive than that, and that maternal impression was only one small aspect of much more widespread occurrences. With regard to maternal impression he gave many examples of each of three different types. It would appear appropriate to begin with a selection of cases from the fifty he set out as evidence of maternal impression. He stated that readers who took the trouble to study the cases in his table (vol.1, table 3-5) would, in all probability, come to share his belief that "mental images in a pregnant woman's mind have sometimes influenced the form of her gestating baby

The three different types of maternal impression referred to are 1) those in which the mother saw the injury or malformation that corresponded with the infant's birth defect. These are the most numerous; 2) those in which she merely heard about the injury, usually hearing it graphically described; and 3) those in which the mother's body itself suffered the wound or was otherwise affected. Where Stevenson, for comparison purposes, supplies results of research as to the incidence of specific deformities similar to the one in question, such results are included here without reference to his sources, though each source is furnished by Stevenson in table 3-2, pp 111-129, and 3-5, and p. 140. Following are 13 cases from the 37 Stevenson placed in the first category. They are typical, neither better or less representative of the remainder of first category cases.

* * *

As reported by Dr. W.F. Montgomery, in 1857, a woman saw a man, probably before she became pregnant, with malformed feet, namely missing the distal third of each foot. When pregnant, the mother had expressed the fear that her child would be affected by her own repugnance to the malformation. The infant was born with the distal third of each foot congenitally absent. Dr. Montgomery stated that he had personally seen the man with the malformations and the baby with the birth defects of booth feet. The infant died a few minutes after birth.

* * *

Also in 1857 the same physician reported that another mother, while 2 months pregnant, saw a baby with one finger congenitally absent, with other fingers 'syndactylous,' meaning fused together. The mother expressed fear that her unborn infant would be so affected and it was claimed also that the mother was "known to be, at all times, very nervous and easily alarmed." The infant was born with one finger congenitally absent and other fingers fused together. The deformities were said to be identical to those of the baby that the mother saw. The incidence of births with absence of one or more fingers is 1 in every 90,000.

* * *

In 1908 Dr. S. Artault reported the case of a woman who repeatedly saw from her first month of pregnancy, a deformed left index finger of a frequent visitor to the family. As the result of an accident the left index finger was thickened and curved so that it resembled a lion's claw. The mother did fear that the infant would be affected. According to Dr. Artault, the mother was described as "being obsessed by the visitor's clawlike fingernail. Every time he came to the house, she did not take her eyes off the guest's fingernail, terrified that her baby would be born with a similar one." When born the infant's left index finger had a clawlike nail exactly resembling that of the visitor.

*   *   *

In 1856, Dr. Z. Pitcher reported the case of a woman 4 months pregnant who saw a heavy trunk fall on the right forearm of her 3 year old son. The infant she later delivered had a congenital absence of the right forearm. It is not known whether she expressed any fear regarding the fetus she was carrying, however at the time of the accident she was reported to be "very much excited." The incidence of births with unilateral absence of a forearm is 1 in every 22,000.

*   *   *

In 1908, Dr. J. Brault reported the case of a pregnant woman who saw over a period of time a young girl with an "enormous macroglossia,' namely, an unusually large tongue, which can sometimes cause cosmetic and functional problems in speaking and eating. She had seen the girl during a prior pregnancy that ended in a miscarriage. The sight had left a strong impression on her. During the pregnancy at issue here, the mother awoke crying after dreaming about the girl with macroglossia, and thereafter was obsessed with the appearance of the child. She remained convinced that her baby would be similarly malformed. The child was born with both macroglossia and lymphangiomata, a benign tumor composed sometime of cavernous lymphatic spaces in the neck or the space below the shoulder through which vessels and nerves enter and leave the upper arm,. It also provides the under-arm sweat gland in humans. More specifically it is the area on the human body directly under the joint where the arm connects to the shoulder, or in short, an armpit. Brault later successfully operated on both of the affected children.

*   *   *

Dr. M.D. Reeve in 1889 reported that a woman 1 month pregnant held the arm of her brother-in-law after he had had his hand amputated following a crushing injury. When the woman's child was delivered it had a congenital absence of the left hand. We are not told whether the woman had expressed any fear of such a consequence, nor do we know which of the brother-in-law's hands had been amputated. The incidence of congenital unilateral absence of a hand is 1 out of 65,000.

* * *

In 1949, in one of the few cases reported in the 20th century, Dr. L.F. Leclerc-Momtmoyen reported that a woman between 1 and 2 months pregnant saw a man who had only one external ear, the other having been cut off by a sword during war. The woman had expressed fear of its possible effect on the unborn baby. Dr. Leclerc-Momtmoyen knew the soldier whose ear had been cut off, and he delivered the baby with the left external ear absent. He did not state which ear of the soldier had been cut off. No figures of incidence have been reported but through 1984, only 7 such individual cases had ever been reported.

* * *

In 1877 Dr. J. Cargill reported that a in the early stages of pregnancy, her 3 year old son had his penis bitten by a dog with the resulting backward bending of the glans and distal urethra. The glans penis is the rounded head of the penis. At birth the glans of the penis is attached to the foreskin. Over time the foreskin begins to separate from the glans until the skin is able to fully retract. The woman was not known to have voiced any fear of effect on the unborn fetus, but the doctor states that the mother was "very much alarmed at the time." The fetus was delivered with the glans penis congenitally bent with the urethra facing backwards. Dr. Cargill saw both the injured child and the baby with the defective penis.

* * *

In 1863 Dr. A.P. Owen reported that a woman in the early stages of pregnancy examined her brother's genitalia after he had had his penis amputated because of a carcinoma. The woman expressed no fear of a similar effect on her unborn child but she claimed that after seeing her brother's operated penis "her mind was increasingly engaged with reflecting and sympathizing upon the forlorn conditioned of her brother." Her newborn child was born with congenital absence of the penis. As of

1898 the incidence was reported as 1 out of 30,000,000 births. Another source states that as of 1951 only 15 such cases had been reported.

*　　*　　*

In 1906 Dr. J. Lacambre reported on a woman 2 months pregnant who saw an operation on an old woman for closure of an enormous umbilical hernia. The pregnant woman had previously given birth to two normal babies. The operation she saw took place in in the patient's home in a village. The pregnant woman, determined to see the operation, sneaked into the house, quietly opened the door of the room where the operation took place and looked at the scene. Overcome by emotion she fled to the garden whereupon she fainted. She did not express any fear that this would affect her unborn child. However the fetus she carried was born with an umbilical hernia that was proportional for its size, to that of the old woman operated on. The incidence of births with umbilical hernia is 1 out of 23, 413.

*　　*　　*

In 1886 Dr. W.J. Swift reported the case of a mother with a 5 year old, and 8 months pregnant with her 2nd child. The left thumb of the 5 year old was crushed by a closing door. The thumb was swollen and there was ecchymosis beneath the nail, that is, a discolored area on the membrane caused by blood seeping into the tissue as a result of the contusion. There was no expression of apprehension on the part of the mother about any effect on the unborn child. However he was born with a swollen right thumb and ecchymosis beneath the nail. The left thumb had a slight ecchymosis beneath the nail. Dr. Swift examined the injured child and the newborn baby. The thumbnails of both children came off within about 24 hours of each other. New and perfect nails then formed on both thumbs.

*　　*　　*

In 1877 Dr. T.D. Saunders reported the case of a woman, who in her 6th month of pregnancy experienced an accident involving her 2 year old son. The child cut his upper lip severely with scissors so that, in effect, a cleft lip had to be repaired. The mother was not known to express any apprehension of any effect on the fetus she was carrying. Upon the birth of the child she was carrying however there was a linear scar on the upper lip at the same site as the scar on the older brother's lip. Dr. Saunders had delivered both babies and he repaired the cut lip of the older boy.

*   *   *

In 1880 Dr. D.W. Ramsey reported the case of a woman who, six months pregnant, saw her husband's left forearm and hand after they had been "terribly lacerated by a cotton gin so that the hand was "hanging down." It is not known that the mother expressed any apprehension with regard to the unborn child, but she was "terribly frightened" at the sight of her husband though she saw him "but a short time." Before and after this pregnancy she gave birth to two normal babies. When the child she was carrying was born however the left hand was badly twisted on the wrist, the first phalangeal bone of the thumb was missing, and the fore and middle fingers were grown together. Dr. Ramsey saw both the injured man and the defective baby.

*   *   *

Each of the above cases involve the expectant mother actually seeing an emotionally unsettling sight. As already mentioned, there are also cases, understandably fewer in number, wherein the expectant mother merely heard of such an unsettling or frightful event. There appears little reason to use more than one of the 3 such cases, described by Stevenson, involving pregnant women who heard of such events, but did not see them themselves. The following case is typical:

In 1908 Dr. H. Lagache reported a case of a woman who had previously given birth to seven entirely normal children. When 6 months pregnant with an 8th child she heard about, but did not see, a man, a neighbor, who lived about 100 meters away, whose left hand was badly mangled in an industrial accident. It necessitated surgery for the amputation of three metacarpal bones and their associated phalanges, that is, sections of the fingers. There was no expression of fear that it would affect the expected baby, and she said in fact that she had not been particularly affected by the news. The mother however was well known to physicians as neurotic and had been diagnosed as having hysteria. The child in utero, when born, had a malformation of the left hand identical to that of the injured man's hand after the operation which then had only the thumb and little finger. The three metacarpal bones were also absent.

According to the doctor, this case was published with more detail than any previous case. The report included photographs of the injured man's hand after the operation and of the child's hand, as well as x-ray photographs of both the affected hands. As mentioned above, the incidence of birth with missing one or more fingers is 1 out of 90,000.

* * *

As stated, Stevenson mentioned a third category of cases, those in which the mother was herself wounded or otherwise affected. Two of three of those cases are summarized here.

In 1898 Dr. C.F. Gardiner reported that a woman over 4 months pregnant had an accidental contact with a sunfish against her leg. The mother was shocked and "for the moment completely unnerved" by it, but later thought little about it. The baby was born with a colored mole at the site where the sunfish had come into contact with the mother's leg. In later years the mole was 12 by 3.5 centimeters. Gardiner published a photograph of the mole. Both the mother, during her pregnancy, and the daughter born with the birthmark had a strong craving to eat sunfish.

* * *

In 1891, Dr. J.W. Ballantyne wrote of a patient who, while 7 months pregnant, was bitten by a dog on the right leg. She did express concern about a possible effect on the child she was carrying. Her wound had been cauterized and had suppurated, that is, filled with pus, afterward. Immediately after the baby's birth "the mother asked anxiously if there was anything the matter with the child's right leg. The baby was in fact born with a "reddish scar" under the right knee.

Chapter 4

Birthmarks Corresponding to Wounds Verified by the memories of Informants

We now move away from cases illustrating maternal impressions. Stevenson also presents 10 cases illustrative of its title, two of which will be summarized here. We will see in these and in cases from following chapters that the similarity of birthmarks on the subject are as close to the wounds of the prior personality as are the corresponding features in cases dealing with maternal impressions.

The first case is that of **Mahmut Ekici** (Ekici), born in July 1923 in Havutlu, a village in southwest, Turkey. It is about six miles south of the major city of Adana, and about 16 miles from the coast of the northern Mediterranean. His parents were Mikail Ekici and his wife Fatma. He was born with a prominent birthmark, which was one factor that led Fatma to believe he was the reincarnation of her first cousin, a man named Mahmut Namik (Namik). His family also lived in Havutlu. Namik was stabbed to death some time before Ekici's birth. She also claimed that Ekici, at age 2 began to talk about the life of Namik.

We turn first to the circumstances of the death of Namik. After the defeat of the Ottoman Turkish Empire during World War I France occupied the area around Adana in 1919. The French withdrew from that area, as required by treaty in 1923 so it was between those dates that Namik was killed. According to Ekici it was in 1921, but there is no other evidence of the exact date of death.

During that period of occupation the Turks engaged in partisan warfare against the French occupiers. Namik became involved with this struggle, but the evidence is not clear as to whether he was a partisan Turk fighting the French, as Ekici suggested, or the victim of a personal quarrel. There is no identified eyewitness to his death and the body was not found until 4 months after his disappearance. Ekici also said that Namik was between 18 and 20 when killed. Ekici's older sister, Emine, about 11 year old at the time, claimed, on what evidence is not known, that another Arab who had become a French citizen was jealous of Namik and killed him while accompanied by French soldiers, who either permitted or assisted in the murder.

There is controversy concerning the location, involving right or left side, where the fatal blow was delivered. Stevenson seems to think it is of some importance. Considering that the body was badly decomposed when found, the uncertainty should not be surprising. Fatma attended the funeral of her cousin, which Stevenson says might have put her in a good position to see the wound, or to hear about it from a reliable first hand informant. More by gestures than words (the conversation was mostly through an interpreter) she indicated that it was on the left side of his abdomen.

Emine, the deceased's sister first stated that her brother, Ekici, had said that he was stabbed on the right side in the area of the liver. But she later remembered that Namik's mother had preserved the shirt he was wearing when killed. The shirt had become an important keepsake to her. Emine said that she had seen the shirt and that it had a hole on it in the part that would have covered the liver. Stevenson acknowledged he could have wished for stronger evidence but considered this to be strong enough to justify a claim that he had independently verified the location of a wound on the prior personality, one that corresponded with the subject's birthmark.

We turn now to Stevenson's description of the birthmark on Ekici. According to his mother, Fatma, it was bleeding and oozing when he was born. She said it discharged for a month, but was contradicted by an older son who said it was for a year. Ermine claimed it was for two months, a compromise ultimately agreed to by Fatma and the older son. Stevenson displays a photograph (R and B, vol. 1Fig 5-3, p 270) taken in October 1967 when Ekici was 44 years old. Stevenson explains that Ekici pulled his shirt up just before the photograph was taken, causing a stretching of the skin in the birthmark area. This, he says, made the mark appear larger in area but less deep beneath the surrounding skin than it was before the skin was stretched. It is nonetheless quite clearly a visible mark.

After the drainage ceased, whether a year, or one or two months, Ekici continued to have some discomfort in this area, continuing into his adulthood. Especially after he had worked hard and become tired, he felt pain there. It was rather deep within the abdomen, and was not limited to the level of the birthmark on the skin.

His stated recollections of his 'other life' were sketchy at best, and included no facts of which others were not already aware. Though these recollections alone would probably not have been considered substantially significant, the existence of the birthmark was not subject to the same objections. Its correlation with the wound on Namik, as its location had been established by independent evidence, lent a degree of significance to the memories that might otherwise never have been deemed sufficient.

*   *   *

We turn next, in this category of wounds verified by memories of informants, to the case of **Aristide Kolotey**, born in Kordaki, Ghana on September 12, 1945 (p 340). His story begins however with what Aristide learned, at age 17, about the death of one, Poepak, referred to as Aristide's uncle, from his father's brother, Immanuel, hence definitely a real uncle. As remembered by Aristide, Poepak's body washed up on shore about a week after he had disappeared. The body had a cut in the middle of the chest, but death was attributed to drowning. It was concluded that he had dived into the water without seeing a rock under the surface. He probably had cut himself badly on the rock, and consequently unable to swim, he drowned.

Aristide told Dr. Stevenson that he thought he had not any imagined memories of Poepak's life when he, Aristide, had been a young child. The only evidence that might, or might not have a bearing, other than the birthmark, was Aristide's statement that he had a "marked phobia of water." He did not learn to swim until age 38, but still did not enjoy being in the water.

Emmanuel saw Poepak's body after it washed ashore, thus providing some second hand verification of a correspondence between a wound on Poepak's chest and a birthmark on Aristide. That mark was seen from birth and extended almost the entire length of the front of his chest and upper abdomen. A photograph (fig 5-18) appears in Stevenson's Volume I. According to Stevenson, the birthmark began 1-2 centimeters to the right of the midline just beneath the right clavicle and ran "inferiorly and medially," or down and across. It extended into the upper abdomen. It had the appearance of an acquired scar. It showed lesser pigmentation than the surrounding skin. It was not hairless. At its widest it was about 3 millimeters wide.

Stevenson comments that Aristide's birthmark was about as obvious as any can be, a fact that causes him to wonder why at an earlier age Aristide was not told of the belief in the family that he was Poepak reincarnated. The question seems answered by facts in Stevenson's narrative of the case. In essence, he was raised as a Christian, and as a boy his father worked in the large city of Accra on the southern coast and was away most of the time. Aristide himself was trained as a biochemist with a doctoral degree and was employed in the area of Washington D.C. He told Stevenson that his Christian upbringing had probably prevented him from learning as much about that as he otherwise would have.

*　　　*　　　*

Chapter 5

Birthmarks Corresponding to Surgical Wounds Verified by Medical Records

The strongest evidence of the wound on the prior personalities are those for which medical records were examined. These, of course, are not subject to the vagaries of human memory, outside influences, or other sources of error that could creep into the narrative of human eyewitnesses. Dr. Stevenson sets out five such cases, of which two are summarized here.

The subject of this first case is **Alan Gamble**, born February 5th 1945 in Hartley Bay, British Columbia, Canada. His parents were Clarence Gamble, a trapper, and his wife, Flora. We start however with the prior personality, Walter Wilson and the circumstances of his death on February 18, 1942, three years before the birth of Alan.

Wilson was born in 1921, lived at Hartley Bay, married at 19, and had been adopted into the Gamble family. He was hence by adoption the brother of Alan's father and the Uncle of Alan.

The only eye witness to the injury from which Walter died was Leonard Davidson, the husband of one of Walter's sisters. Sometime in early February 1942, Walter and Leonard left in a small fishing boat. A skiff was being towed. At one point as the boat neared a shore, Walter spotted a mink running on shore near the water. Walter's 12 gauge shotgun was in the skiff with the barrel pointed toward him. From the stern of the boat he pulled the boat toward him and reached for the gun with his left hand. It slipped and the butt fell and struck a board of the skiff.

The gun went off and shot Walter through the left hand. Bone protruded through the wound and there was severe hemorrhaging. Leonard applied a tourniquet to Walter's upper arm. It was several hours before the boat could reach a larger boat in which he was taken to Prince Rupert. Sometime during that journey Walter lost consciousness and apparently never regained it. It was night when he was admitted to the Prince Rupert Regional Hospital. Leonard had been unaware that a tourniquet must be loosened periodically to permit circulation to the parts below it. It had, in fact, been left on the arm continuously for 10 hours, resulting in gangrene.

The upper third of the forearm was amputated, but the wound became infected, and as mentioned above, Walter died on February 18th 1942. Stevenson could not obtain the record of Walter's admission to the hospital as all records prior to 1959 had been destroyed by the hospital. The death registration, however, contained the following:

> **Cause:** *Gunshot wound left forearm, aggravated by a tourniquet in the middle of the left arm, left on for 10n hours.*

> **Contributory Causes:** *Septic infection in the wound and amputated stump, and shock.*

> **Operation:** *Amputation of left forearm, upper third, on account of gangrene.*

> The death certificate attributed the cause of death to "accidental injury by firearms."

Alan's mother, Flora, told Dr. Stevenson that Alan, when young, talked a lot about the 'previous life," though she could recall few details. Both the sight of shotgun shells and the illustration of them cause him to talk. He would say that they were the cause of his having a sore hand. Once he showed fear upon seeing shotgun shells. The family, realizing that his memories were troubling him hid the shotgun shells from his view.

There were several other matters, minor and otherwise that evidenced his memories of another life.

More important from our viewpoint is the description of birthmarks on Alan. There was one on the palm of the left hand, which Stevenson considered the entry wound. A photograph of it, taken when Alan was 3 ½ years old, was published by Stevenson. It was about 3 millimeters long and 1 millimeter wide. It was depressed by .5 millimeter below the surrounding skin and looked like a small scar, though it did not differ in pigmentation from it.

There was also a birthmark on the dorsal portion on or toward the upper back side of the left wrist, likewise illustrated by photograph, and which Stevenson determined to be the exit wound. It was significantly larger than the one on the palm. It was dark purple-red in color, rounded in shape, about 8 millimeters in diameter. Neither mark had changed position or size relative to the other or to surrounding tissue since Alan's birth.

Alan's older sister, Betty, said that his left hand was swollen at birth, though the mother, Flora, claimed it began a day or two after birth. The swelling of the arm began at the site of the amputation of Walter's arm, which was about one third of the way down the forearm from the elbow. It then continued down toward the hand until the entire lower part of the arm was swollen. Alan's parents took him to the hospital for treatment of this condition where he remained 2 or 3 weeks and recovered.

Even beyond childhood, Alan endured intermittent pain and swelling of the left arm, and once had to leave a job because of it. Flora thought he had pain at least until 1977 when Stevenson met her. In 1979 Alan denied ever having pain in the left hand. Stevenson found that the left hand was in fact functionally normal. He also noted that the amputated part of Walter's arm corresponded to the part of Alan's arm that swelled soon after his birth.

*   *   *

The subject of the second, and last, case in this chapter is **Henry Demmert III**. We deal here with three generations of Demmerts: Henry Demmert III, the subject of this case was born in Juneau, Alaska on October 5, 1968. His parents separated soon after his birth and Henry was adopted by his maternal grandparents, Henry Demmert, Sr. and his wife, Gertrude. It was already believed that the child was the reincarnation of a predeceased son of Henry Demmert, Sr., namely Henry Demmert Jr who had died in 1957. The mother of that son was by Henry Sr.'s first wife, Muriel. The belief about reincarnation was based in part on the existence of a birthmark on the child, Henry III. Henry Jr., his predeceased father, was the presumptive prior personality. So we should look first at the life of Henry Jr. and the wound in question.

Henry Sr. married Gertrude in 1932 when Henry Jr., his son, was 2 years old. Jr. was, hence, raised by is stepmother. Jr. married, and at the time of his death he, his wife, and their two children were living in Juneau. He was generally known to be involved in bouts of drinking. On March 6th 1957 he went to a party at a private home where there was heavy drinking that lasted through the night. At about 5:00 A.M. Jr. was stabbed in the heart. He was rushed to the hospital where he died at 6:45 A.M.

Though Jr.'s parents considered his death to be a case of murder, after inquiries the police made no arrest. A coroner's jury concluded that death resulted from an "accidental self-inflicted wound." Stevenson commented that the wound was one not likely to be self-inflicted except in case of deliberate suicide, which was not at issue here. He suggested that the knife could have been pushed into the body, even unintentionally by the opponent. He suggested further that Tlingits sometimes fail to "resolve questions of causations," preferring "turbidity" as the best means of allowing animosities to subside.

More important is the wound itself. The autopsy report stated as the primary cause of death, "laceration of the left lung and heart causing exsanguination." As "Major Findings of Operation," namely the autopsy, the following words were added: "Knife wound pierced the heart." This was the fatal wound of Henry Jr, What about the birthmark on Henry III?

Once again we have a photograph of it taken in 1978. As described by Stevenson the mark was inferior and slightly lateral to the left nipple. It was approximately at the level of the 6th rib. "It was a hyperpigmented macule approximately 3 centimeters long and 8 millimeter s wide." A macule is a patch of skin that is altered in color but usually not elevated. "The medial end of the birthmark was slightly more pointed than the lateral end. The birthmark was not elevated, but may have been fractionally depressed below the surrounding skin."

The birthmark was noted shortly after the birth of Henry III. Both grandparents, Henry Sr. and Gertrude said that it corresponded in location with the fatal wound on Henry Jr. Only Henry Sr. was actually interviewed by Stevenson, Gertrude having died before the interview. Stevenson stated that "a knife entering the chest at the site of the birthmark, or near it, would penetrate the heart if directed medially and upward." At the birth of Henry III the mark did not bleed or ooze. Henry Sr. said that it looked like a 'scratch mark." When Henry III was about 2 years old he made the only two statements about the previous life that he is known to have uttered. He said of his birthmark, that he "got hurt there." He also added that it happened when he "was big." We have seen the same expression, "When I was big," by very young children, used in two cases summarized in *Our Quantum World and Reincarnation.*

Chapter 6

How Does It Happen?

What is transpiring physically? The starting point for such an inquiry must necessarily be the same as that expressed by Drs. Ian Stevenson and his successor at the University of Virginia, Dr. Jim B. Tucker. Tucker was researcher of some of the cases summarized by me in earlier publications. In his *Life before Life* he writes that mental factors can produce general changes in the body, such as stress, weakening the immune system's ability to prevent infection. Hopelessness, he points out, increases the risk of heart attack or cancer (p69 of *Life before Life*). We must agree with his starting point as the role of psychology, mental outlook and emotion will here be shown to play a key role in precipitating the phenomena described in Chapters 3, 4, and 5. As we shall see a bit later, Dr. Stevenson is also in agreement.

As we delve further into the matter of causation of the physical phenomena we have just seen however, I depart significantly from one important aspect of the thinking of Drs. Tucker and Stevenson. The issue involved is neatly presented by author and psychologist Nessa Carey in her *The Epigenetic Revolution*. (p 234). She begins with an observation identical with that of Stevenson and Tucker: "An abusive or neglectful environment when young is clearly a major risk factor for the development of later neuropsychiatric disorders." As adults, she continues, they are at significantly higher risk of other conditions including schizophrenia, eating disorders, personality disorders, bipolar disease, generalized anxiety and abusers of drugs or alcohol.

She then asks why events that lasted for a relatively short period in the life of a young person have adverse consequences decades later.

She explains that a religious person may prefer to invoke the soul; a Freudian therapist may invoke the psyche. She rejects both. "Both of these refer to a theoretical construct that has no defined physical basis." Those words are clear enough, but she refines it further in language equally as penetrating. Scientists, she writes, prefer to look for a mechanism that has a physical foundation. She finds that more satisfying than "defaulting to a scenario" in which something is "assumed, to be a part of us, without having any physical existence." A scientist, she says, would want to probe the molecular events that underlie the psychological damage.

The "molecular events." She speaks of the resistance to this approach from other disciplines, "which work within different conceptual frameworks," something that, she says, puzzles her.

It is from this point that I take leave from the explanations of the two explorers of this newly discovered realm of the human condition. Drs. Stevenson and Tucker both believe, I sense, though possibly erroneously, that the souls of the deceased persons, whom they call, as shall I, the "prior personalities" to be the vehicle for the transfer of memories. It must be said in behalf of these two scholars that neither of them are inclined to make an issue over this controversy that cuts so deeply on all sides, so much so that I cannot be certain that my reading of their minds is correct. They are content, like the scientists and scholars they are, to present us with facts, but their beliefs seem to me to leak through. If I am incorrect, then I feel that either or both of them have decided to leave this mystery to the vast field of the unknown. I would not be in agreement with either position, or with leaving the mechanism to the unknown. They have reckoned, perhaps justifiably, without consideration of the relatively new subject of epigenesis, a matter to which we will turn a bit later. If that is not the right answer, the right one will eventually be found by science. At this point, I believe epigenesis to be the best candidate

But before taking leave of their positions, I would like to salvage well expressed opinions of each on another, and easier aspect of this particular subject. Both agree that the mental element is a crucial ingredient of the serious and often permanent effect of mind over matter, a physical disability of oneself or of others resulting from emotional disturbances beyond control.

Stevenson introduces his discussion with a thorough treatment of the subject of *stigmata*. It refers to the wounds inflicted upon Jesus at the crucifixion. Persons later developing wounds, apparently spontaneously, in similar location and appearance to those on Jesus were said to have received the 'stigmata.' The first and best known was St. Francis of Assisi.

As of 1997 there were at least 350 cases of stigmata reported, though, according to Stevenson, not all were of equal authenticity or interest. There have been a small number of fraudulent cases exposed, including some who pierced their own skins, apparently to share in the suffering of Jesus. Some of them might have been in a trance. There are others however in which the testimony shows "beyond doubt" that the lesions were not created artificially (R and B, Vol 1, p 34).

These include persons of saintliness, and others who fall short of any such quality. Stevenson mentions three cases of particular relevancy to our focus here. In each, stigmatists developed changes in the tissues of the wrists that were believed to be incurred by Jesus from the ropes binding him when first arrested. The cases are relevant to cases in recent times and reported by Stevenson, wherein there exist descriptions of ropelike marks on a psychiatric patient who relived the experience of being bound by ropes, and on a few children who remembered the lives of deceased persons who had been bound with ropes.

Stevenson concludes that persons who develop stigmata either must be willing "to concentrate singlemindedly for weeks, months, or longer on the wounds of Jesus, or they must have the quality of unusual impressionability." He then turns to other examples of imitative bodily changes, closer to our area of interest. He starts with a generally recognized phenomenon of the husband experiencing some of his pregnant wife's symptoms such as nausea, vomiting and abdominal pain. This he follows with five documented examples that are even closer to our target.

First, During the French invasion of Russia in 1812, a Russian civilian was by chance trapped in an area where he was forced to witness, with deadly fear, a sword fight between a Cossack and a French soldier. When the civilian finally went home he found on his own body, wounds similar to those the Cossack had inflicted on the French soldier. The wounds bled, and were treated medically. He had not suffered from swords or any other weapon or instrument.

In 1923 there is a report of a young woman whose brother, a soldier, was made to run a gauntlet, involving being beaten on the back by other soldiers. The sister did not observe the ordeal, but was seen to get into a state of excitement at the time of the punishment. She whined and moaned, then fainted. Her back showed signs of having been beaten and was bleeding.

In 1872, the affected person reported that she observed a child coming out through an iron gate, and that the gate seemed likely to close on him with such force as to crush his ankle. That did not happen, but the person reporting claimed that she could not move to rescue him because of the intense pain that came on in her ankle, corresponding to the one on the boy that she thought would be injured. She was certain she did not strain or sprain it. The quarter of mile walk home was very laborious, and she then found a red circle around the ankle. By morning the whole foot was inflamed and she was obliged to remain in bed for several days.

In 1959 a reported case described a young mother who developed blisters on her right upper arm following vaccinations of each of her four children. She had previously developed one when she first became pregnant. She later, during a period of stress, developed a blister on the same arm, but 2 centimeters distant from the earlier ones. On three of the six occasions herpes simplex virus was cultured from the blisters. In each the mother felt slightly indisposed before a blister developed. It took a few days to develop and lasted a few days when it did. Before the first pregnancy the mother had not suffered from any skin disease, or had blisters on her skin or mucous membrane.

The following and last case contains a thoughtful and professional opinion about causation.

Published In 1853, it involved a lady who was watching her little child playing, and saw a heavy window sash fall on his hand, cutting off three of his fingers. Her fright and distress disabled her from rendering assistance, but a doctor soon reached the scene. After dressing the wounds he turned to the mother who was seated, moaning, and complaining of pain in her hand. The doctor found three of her fingers, corresponding to those injured on the child, to be swollen and inflamed, though they had not ailed in any way prior to the accident. In the next 24 hours incisions were made on the mother, pus evacuated, sloughs discharged, and the wounds ultimately healed.

The doctor concluded that "the mother's emotion was directed, by observation of the parts injured, upon the corresponding parts of her own system, there working a change in the circulation or nutrition, sufficient to excite acute inflammatory action" (R and B, vol. 1, p55). We will see later in our discussion of epigenesis the high degree of prescience that may be contained in that opinion.

Stevenson makes clear that his belief with regard to requirements of the stigmatists applies equally to other examples of the same phenomena, such as those in the above 5 examples, and in the vast majority of his reported cases. They must either be willing to concentrate, or unable to avoid concentration singlemindedly for weeks, months, or longer on the wounds of others, or they must have the quality of unusual impressionability.

What about Dr. Jim Tucker? He makes clear that he shares the belief of Dr. Stevenson. He adds to the examples of Stevenson the subject of hypnosis and the physical changes it can bring about. He mentions the production of blisters on subjects by telling them they were being burned and then touching them with a cool object. At times when the hypnotist used a recognizable symbol, subsequent wounds were in that shape. Also suggestions under hypnosis can produce not only the suggestion of thirst, but also changes in the kidneys that occur during dehydration, also changes in heat rate and control of bleeding.

He also makes clear his opinion. "The mind can produce changes in the body that, given our present state of knowledge we are unable to explain." By "mind" Tucker writes, he does not necessarily mean the brain. He means the "world of thoughts, or consciousness that exists in the brain" (*Life before Life*, p.69).

His explanation that follows that thought, insofar as it goes, is a splendid summary of our thesis here. It will be quoted in this paper in full in chapter 11.

PART II

Epigenesis

Chapter 7

The Underlying genetics

Epigenesis is so tightly tied into the field of genetics that it is impossible to explain what it is, or how it functions insofar as that is presently known, without some understanding of genetics. That understanding should include at least the following basics, many of which will be known by many readers. To such readers, my apologies, and for them the following few paragraphs of such basics can certainly be skipped without significant loss to an understanding of epigenesis which follows it.

There are @50 trillion cells in the human body. Each cell contains 3 billion 'base pair' of DNA. A gene is part of the 3 billion base pair, which is made up of two strings, side by side, but twisting around each other in a spiral-like configuration. The four bases on each string join with each other chemically in certain fixed combinations and hold the two strings together.

DNA contains the blueprint for the genetic makeup of the individual. This blueprint, or code, contains only four letters: ACGT, for adenine, cytosine, guanine and thymine. All are organic compounds, and they encode, in a variety of combinations, for molecules known as amino acids, of which there are 20. The twenty, in turn, are the building blocks for proteins. The proteins deliver oxygen through the blood stream and extract nutrients from food for delivery to the various parts of the body. The DNA are in much smaller bundles called chromosomes.

Each human has between 20,000 and 24,000 genes. Every pair of joined bases is known as a base pair. A typical gene may consist of approximately 3,000 base pairs. Over 99% of the three billion bases, are the same in all people.

Of the 20,000 or more genes only about 2% of them do in fact code for proteins. Just about 25 years ago the other 98% was still called "junk DNA." No one knew what they were for, and were written off as left over from a time in evolution when they served some purpose, but now no longer have any purpose.

But even then, there was research being done that ultimately showed that this junk DNA controlled whether the genes, that code for proteins, would be turned on or off, and to what degree they would function, if at all.

*   *   *

As recently as 1997, it could be written in a scholarly journal *(Current Directions in Psychological Science 6.4, August 1997, 106-110)* that "the importance of genetics for various aspects of human behavior hardly is a matter for discussion any longer." This passage, and the entire article are both noteworthy, not so much for what they say, but for what is left unsaid, namely, the role of epigenetics.

Acceptance of the notion of epigenetics, like many new ideas and discoveries in science was met with stubborn resistance. It was mostly ignored for decades. It ran counter to long accepted, well entrenched dogma that had taken root in the field of genetics. Our focus in this chapter however will be on the subject of epigenesis and its possible role in one of the more mystifying aspects of the phenomenon of children who remember other lives. Examples of that mystifying aspect, marks and defects in the subject person similar to those inflicted upon the prior personality, has been the subject of prior chapters of this essay.

I believe, as set forth in two prior writings, that the aspect of quantum physics known as entanglement is a possible explanation for the purely mental part of the life of the prior personality, that is, memories and associated emotions that devolve upon the subject fetus or infant. However the appearance of birthmarks and birth defects, identical, or very similar, to wounds or disfigurements suffered by the prior personality, during his or her lifetime, is not purely mental. On first blush it is not mental at all, but very, very physical.

I believe that just as a physical, biological process, entanglement, may be the solution to the memories of others in young children, a physical, biological process, epigenesis, may point to a solution of the birthmarks or defects in question. That was the reason, as stated, for the basics outlined above.

It must be stressed at this point, for it is now generally accepted: The epigene is, unlike the genome, sensitive to the outside world, to the environment. Change will come quicker through the epigene, but not as permanent, though usually long lasting enough to affect another generation or two, or three. Further, through a process called canalization, the path taken by epigenetics may serve as a guide for the path to be taken by mutated genes.

The question arises: What is the molecular agent, the epigenetic mechanism, if any, that affects the status of the gene: active or inactive, and if inactive, to what degree?

Chapter 8

Mythelation, and the Discovery of Epigenetics

David S. Moor, a development cognitive neuroscientist sums up the question posed by the last chapter: "There are several different epigenetic mechanisms that can silence or activate DNA, and most of these affect the histones, briefly described below. At least one mechanism interacts with DNA directly. The best understood mechanism that interacts with DNA directly is known as 'DNA methylation'. . . a process by which a molecule called a 'methyl group' gets attached to a DNA strand."

Moore further explains that once the methyl group is attached to the section of a DNA strand, that section closes up and is physically blocked from the 'machinery' that would otherwise react with it, and the DNA strand would not be transcribed. If a DNA strand takes on other methyl groups, it further reduces expression of genes on that strand. If a DNA strand is stripped of some of its methyl groups, chances are increased that genes on that strand will be expressed. So in essence, DNA generally silences genes. (*Developing Genome*, pp 40-43).

A methyl generally is a piece of a molecule with the general formula (CH_3). It is a fragment of a methane molecule (CH_4); The -yl ending means a fragment of an alkane formed by removing a hydrogen atom from the methane. (*Epigenetics Revolution*, 55-57; 83-90; 106-110; 254-266; 267).

Mythelation is not the only agent that effects expression of genes, and consequently epigenesis. DNA is closely associated with proteins, and of most particular importance are those with proteins called 'histones.'

Histones are small proteins of five major types. The human genome, as has been mentioned, contains approximately 3 billion base pairs of DNA packaged into 23 chromosomes.. The human body contains about 50 trillion cells—which works out to 100 trillion meters of DNA per human. Consequently each of us has enough DNA to stretch from here to the Sun and back more than 300 times, or around Earth's equator 2.5 million times! Yet two meters of this linear DNA can be compacted in the cellnucleus of roughly 10 micrometers in diameter, a micrometer being .0001 or 1/1000 of a centimeter.

How? Certain proteins, namely the histones, do the job. The resulting DNA-protein complex is called chromatin. DNA is negatively charged so the positively charged histones bind with DNA very tightly.

Consider the closeness, the binding in fact, of the DNA in our bodies to the proteins in the histones in the nucleus of our 50 trillion cells. Consider that science has established that nucleosomes are a "general gene repressor. Consider further that scientists believe that nucleosomes carry epigenetically inherited information. It all seems to offer support for the possibility that the nucleosome is, in part, responsible for the differences, even in identical twins. And, in part, for the birthmarks and birth defects described by Dr. Stevenson and other researchers.

There are other important functions of histones. They also make modifications to the genes, but usually less significant ones than do the methylated DNA.

The term *'epigenetic marks'* is sometimes used (*Developing Genome*, pp. 38-40). It refers to both the DNA methylation modifications and histone acetylation modifications. The 'acetylation' refers to atoms known as an acetyl group. It physically attaches itself to histones just as a methyl group attaches itself to a DNA strand.

Histones can also be methylated. (*Developing Genome*, pp. 72-73). What are the differences among the different modifications? DNA methylation, as we have just seen is usually associated with gene silencing; histone acetylation will cause a section of DNA to open up and renders it accessible. DNA methylation is very stable, more so than any histone modification, but it is demethylated at least twice in the life cycle, including once right after conception. It is nonetheless transmitted faithfully during cell division to all of the daughter cells. Effects of histone methylation are not so clear. It can silence, or activate genes in close proximity. Its effects, as far as presently known, are not so certain.

Why was the notion of epigenesis so slow to be accepted? One factor was certain dogma that was part of the "new Darwinism". It can perhaps be summed up in few words that tell us, not how biologic inheritance works, rather than how it doesn't. There was through most of the 20th century, a firm belief among biologists and other scientists that there could be no "inheritance of acquired characteristics." This meant that no matter what the behavioral or physical characteristics developed by a person during his or her lifetime they would have no effect on characteristics of the offspring. The strong arms of the father, or the talents of either parent, developed only through lifestyles would have no effect on the offspring.

There was according to the thinking throughout that century, strict adherence to what was known as the "Weisman genetic chastity belt." The Weisman referred to is August Weisman, a German biologist of high renown. He died in 1914 at the age of 80.

Weisman had studied in the laboratory the phenomena of the cell, and specifically of the chromosomes within it. It was later found that the DNA, discovered in the mid-20th century, was contained within the chromosomes. In probing the question as to whether acquired characters could be inherited, Weismann came to develop his theory of the plasm as part of the germ cells. He decided that acquired characters could affect the soma, or body, but could not be inherited, for to become heritable, the changes would have to affect the germ plasm itself, for which no means were available except for mutation within those cells. To claim validity for the inheritance of acquired characteristics became anathema.

According to his studies and writings there could be no penetration of the germ cells, namely the egg or sperm, by any changes or mutations of the 'soma' cells, the cells of the rest of the body. Each mature germ cell has a single set of 23 chromosomes containing half the usual amount of DNA and half the usual number of genes existing in most other cells in the human body. When they combine, the resulting single cell is followed by the cells of the embryo, then by the fetus. They have the usual two sets of chromosomes, 23 in each, one from each parent.

The mature germ cells are each called gametes and the union of the two is a single cell, the one referred to above, called a zygote. The single zygote cell then begins a series of divisions, forming 2, 4, 8, 16, etc., and ultimately into trillions of cells.

After four to six days - before implantation in the uterus - this mass of cells is called a blastocyst. The blastocyst consists of an inner cell mass and an outer cell mass. The outer cell mass becomes part of the placenta, the organ that connects the developing fetus to the uterine wall to allow nutrient uptake. It also keeps the unborn baby's blood supply separate from the mother's.

The inner cell mass is the group of cells that will differentiate to become all the structures of an adult organism. This latter mass is the source of embryonic stem cells – 'totipotent' cells, which means cells with total potential to develop into any cell in the body, for instance, liver, skin, or neurons and a plethora of others. For our later purposes here, in understanding the process of epigenesis, the nature of the stem cell is of prime importance.

In a normal pregnancy, the blastocyst stage continues until implantation of the embryo in the uterus, at which point the embryo is referred to as a fetus. This usually occurs by the end of the 10th week of gestation after all major organs of the body have been created.

It appears that another important contributor to the resistance to epigenetics was its surface similarity to 'Lamarckism.' Jean-Baptiste de Monet de Lamarck died in 1832, twenty seven years before Darwin's *The Origin of Species*, and two years before Weisman's birth. He, like almost everyone else at the time, did in fact believe in the inheritance of acquired characteristics. What earned the unwarranted ridicule, that often follows him even today, was the mechanism he advanced as the driving force of change, namely, the use and disuse of parts of the body. His ideas came closer to the more scientifically proven study of epigenetics than seems ever to be acknowledged. But its very closeness may have helped to ignite the long delay in resistance to that modern study.

Weisman was a firm believer and supporter of Darwin's theory of evolution. But Darwin, in his classic *The Origin of Species* claimed also to believe the idea of the inheritance of acquired characteristics, together with his own 'survival of the fittest.' That latter term refers to changes that occurred in the lifeline of all species and sub-species, including humans, and holds that in the struggle to survive, the most fit will be successful and procreate. However it was Weisman himself, in an 1889 publication, who to the satisfaction of almost all of the profession, attacked the idea of inheritance of acquired characteristics.

Included in his study was his experiment involving cutting off the tails of 5 successive generations of mice, later generations still being born with tails, apparently an unnecessary exercise. It was he himself who pointed out that despite repeated circumcision of over a hundred generations of Jewish men, successive male generations continue to be born with foreskin (Developing Genome, p. 148).

What, then, could cause change in the germ cells that could result in Darwinian evolution?

It would be helpful to understand that Darwin, like Lamarck, knew nothing about genes. Darwin's Origin of Species was published in 1859. Just two years earlier, Augustinian friar Gregor Johann Mendel, in the course of his avocation involving plants, discovered the existence of something in them, which were later, in the early 20th century, first called genes. Mendel had published a paper describing his find in 1857, but his findings were ignored until about 1900, when its relevance to Darwinian evolution was discovered and recognized. That was too late to be of any use to Weisman whose work ended before the 20th century though he lived until 1914. It was not too late for a Danish biologist, Wilhelm Johannsen who coined the word 'gene' to describe in1909 what Mendel had discovered. Nor was it too late for Conrad Waddington, destined to one of the original developers of the theory of epigenetics, who was born in1905, and who lived for 70 years.

Natural Selection, the mainstay of Darwin's theory of evolution, was until relatively recently believed to result mostly from purely accidental and random changes in the genes of procreation. That included mutations in the course of transcription of genes, or changes in gene frequencies, or combinations which result from random genetic drift, or the result of intermarriage between populations. Mutations were by far the most frequent occurrences. Among the drawbacks of reliance on such chance happenings was the fact that they occurred in individuals, not populations. Some other means of its culture-wide spread would presumably have to be found.

Other than those few pesky matters, it is all very understandable and reasonably simple. Each one of these methods would undoubtedly be very slow and gradual requiring many eons, which is the way in which we think concerning changes from one species to another. Hundreds of thousands or hundreds of millions of years separate fully evolved Australopithecines, *Homo habilis*, *Homo erectus*, Neanderthals, etc. Over three billion years passed between the first primitive life forms and our own *Homo sapiens*.

It all fit together very well. Most changes resulted from accidents, chance. Case closed. Until, that is, we look at the finer details of the body and inborn instincts of each of the millions of species of life on Earth.

Chapter 9

Flies in the Ointment

We can mention relatively small matters such as finger nails, toe nails, eyebrows etc. before looking at some more serious ones, and wonder if each represents a chance error in cell division, a mutation. Or did we grow them because we needed them? Granted, we may have inherited them, and other small matters, from our early mammalian ancestors. No matter. Whenever and to whomever each appeared the question still lingers: was it merely through some errors in transcription?

Even more glaring, the dentition, the teeth, of ourselves, and each of our other animal relatives, molars, incisors etc., seem exquisitely suited to our respective environments and the available food stuff. Was each such pairing an accident? Chance? A mistake in transcription? Why did they (we) all have teeth that fit so nicely with the required digestion of convenient food? What came first, the food change or the new teeth?

There have, in the past, been many attempts by scientists devoted, justifiably, to Darwinian evolution, but apparently purposely oblivious to any thought of modification that smacks of Lamarckism. But the evidence that something was missing was everywhere and too obvious to ignore.

One of the historical episodes that played a key role in precipitating interest in inheritance though other than genetic means was the Dutch Hunger Winter. A German blockade in the fall of 1944 aggravated by a bitterly cold winter combined to create a catastrophic decrease in the available food for the population of Holland. Its population attempted to survive on about 50% of its normal caloric consumption. People scrounged for anything edible, and not so edible, such as grass and scraps of furniture. Over 20,000 people died from hunger. Liberation came in the spring of 1945, followed by normal diets.

Some very surprising results were later found to have followed. If a mother was well fed around the time of conception and malnourished only for the last few months of pregnancy, the baby was often born small. Surprisingly it was later found that babies thus born small stayed small all their lives despite availability of food. Their bodies never got over the early malnutrition. Even more surprisingly the children whose mothers had been malnourished only early (as opposed to late) in pregnancy had higher obesity rates than normal, though they had seemed healthy at birth. As adults, these offspring had other health problems as well, including mental health, including greater susceptibility to schizophrenia. Further, the grandchildren of women malnourished during the first three months of pregnancy showed the same effects (*Epigenetics Revolution,* p. 103).

This was all contrary to the accepted biologic teaching of the time. 'Inheritance of acquired characteristics' was not recognized. Whatever happened to the parents before conception could have no effect on the child unless it involved a mutation in the sperm of the father, or the egg of the mother before or during pregnancy. Hence, evolutionary change is very slow, depending on mutations in the pertinent DNA.

The explanations offered by biologists for these anomalous effects of the 'hunger winter' were numerous. Epigenesis was one of a number of theories. Others include 'Lamarckism,' a theory that received less serious attention than it might have, one that ran contrary to the accepted dogma of the time and advanced the idea of inheritance of acquired characteristics. Other possible theories advanced include 'transgenerational inheritance of non-genetic, features '(*Epigenetics Revolution,* p. 127) and 'transgenerational effect,' meaning that a mother of the group in question will pass more nutrients than normal across the placenta to her fetus (103). Neither of these processes, though interesting, have anything to do with epigenesis, and require no further discussion here.

Though many instances of apparent epigenetics may also be ascribed to transgenerational effects or features, namely Lamarckian, others cannot, leaving the efficacy of epigenesis in some cases proved beyond doubt. One such case involved males in northern Sweden (Epigenetics Revolution, p.104-5). In the late 19th and early 20th centuries severe food shortages caused by failed harvests, were interspersed with periods of plenty. Intergenerational studies ruled out intrauterine environment and cytoplasmic effects. The researchers concluded that it seemed "reasonable to hypothesize that the transgenerational consequences of food availability in the grandparental, the third, generation were mediated via epigenetics."

One writer, among many, who could not accept what was now called Neo-Darwinism, namely strict adherence to the Weisman 'chastity' rule, was Arthur Koestler, especially in his 1976 volume (*The Ghost in the Machine*, p.158).

Among the cases he cited, as evidence of "assimilated characteristics," were the callosities on the African warthog's wrists and forelegs on which they leaned, the similar growth on camels' knees, and the two bulbous thickenings on the ostrich's undercarriage on which they squatted. These are found in the embryos and are inherited characteristics. Did they evolve by chance mutations just where the animals needed them? Koestler doubted it. He saw them as examples of inheritance of acquired characteristics, in short, Lamarckian.

There were other disquieting matters causing the scientific interest to quicken in the late 20th century, such as the following reports, both having been reported in 1997. The first was by David N. Reznick, published in *Science*, vol. 275, entitled "Evaluation of the rate of evolution in natural population of guppies."

In this case, lizards from one of the Bahama Islands were moved to 14 other islands in that group. The vegetation in each island differed in varying degrees from that of the native habitat. Generally short hind limbs on the lizards are better adapted to thin twigs of bushes like those in the new habitats, as opposed to sturdier branches of trees like those on the trees of the native island. Over a period of ten to fourteen years the length of the hind limbs decreased in each population, the decrease being correlated with the 'perch diameter' (how far they could reach while standing on their hind legs) of the vegetation of the new island.

Many, early on, including C.H. Waddington, in his 1975 volume (*Evolution of an Evolutionist*) found such things too rapid for any evolutionary change limited to the rules of Neo-Darwinism. The substance of the article had been known within the scientific community well before the publication in *Science.*

In another case, published in the same volume of *Science* by Virginia Morell, entitled "Predator-free guppies take an evolutionary leap forward," the researchers removed a small population of guppies from a waterfall pool in Trinidad, one that teemed with predators. The guppies were relocated to a pool upstream inhabited by only one predator. In four years, seven generations, the behavior and size of the guppies changed to become like those native to the safer environment. They grew larger, lived longer (presumably even those who escaped the predators) and had fewer litters with fewer and larger offspring per liter. These were all matters that had been determined to be genetically controlled, but the changes were deemed thousands of times faster than would have been required for genetic changes.

Do such changes occur in humans? Over thirty years earlier, a zoologist, Wood Jones, had noted that many Asians squat with their feet flat on the ground, though Australian Aborigines squat with their feet tucked under their buttocks. How do they do it? Jones pointed out that the Asians have facets in the bone structure at the joint of the leg and foot that permit them to squat comfortably in their preferred position, and that the Australian Aborigines have different facets that accommodate their own method of squatting. Jones also noted that most of those who usually sit in chairs do not have such facets (Habit and Heritage, p. 16).

Jones noted also that these boney characteristics exist in the embryos and very young children before they develop their squatting habits, but not in the embryos or newborns of people who use chairs. He commented that "These anatomical features must have arisen very recently in evolutionary terms." Many, including Koestler noted the thickened skin on the soles of babies' feet even before they are born. But this not the way evolution, Darwinian evolution in particular, was supposed to happen

We come to Conrad Hal Waddington. His life spanned much of the 20th century (1905-1975). It was he who introduced the term 'epigenetics' in the early 1940s. That was also the year of his publication of an insightful model for evolution in his 1942 paper, *Canalization of Development and Inheritance of Acquired Characters.* In it he proposed that an unknown mechanism exists that conceals phenotypic variation until the organism is stressed.

The term 'phenotypic' refers to the composite of an organism's observable characteristics or traits, such as its physical build, development, biochemical or physiological traits, and behavior. The observable characteristics of an individual are said to result from the interaction of its genotype with the environment.

He was the first to make widespread use of the term 'assimilated characteristics,' a term first adopted by Arthur Koestler as mentioned above in describing Koestler's own work. Waddington's use of the term resulted in large part from his experiments with fruit flies. (*Evolution of an Evolutionist* pp. 59-95; *Evolutionary Adaptation*. 175-188)

Those experiments show clearly why he spoke of "an unknown mechanism that conceals phenotypic variation until the organism is stressed. Fruit flies have 'anal papillae' which contribute to pressure of the fluids in the flies' bodies. Waddington increased the salt content in the medium in which they lived. This caused the papillae of about 40% of them to undergo certain modifications, including shorter tails. He continued to breed the flies with the modifications, and after 21 generations, the changes persisted in later generations even when bred in the normal medium. Hence the change came from an outside force. To Waddington's perception the outside force precipitated an adaptive change with genetic consequences.

Waddington subjected other fruit flies, usually born with cross veins in their wings, to heat shock during embryonic development. This caused some flies to develop wings with no, or fewer, cross veins. Their offspring for several generations, though not subjected to heat shock, also had no cross veins.

In another experiment the eggs of normal fruit flies were treated with ether-vapor shortly after they were laid. The result was fruit flies with four wings rather than the usual two. Following 20 generations of this treatment on their offspring a stock of flies with four wings were produced in high numbers. Further offspring were then born with four wings even without the ether-vapor treatment.

Waddington concluded that the changes, though heritable, were not Lamarckian. He believed that they occurred through the selection or activation of genes that already existed and the production of new genetic combinations.

Waddington defined epigenetics as "the branch of biology which studies the causal interactions between genes and their products which bring the phenotype into being." In the original sense of this definition, epigenetics referred to all molecular pathways modulating the expression of a genotype into a particular phenotype. In the 1950s he defined it as a stable, heritable trait that can result from changes in the chromosomes without alterations in the DNA sequence.

In more recent times, epigenetics has been defined as "the study of changes in gene function that are heritable and that do not entail a change in DNA sequence." Thus, what was basically his definition was adopted at a meeting in December 2008 of five specialists hosted by the Banbury Conference Center and Cold Spring Harbor Laboratory. The predominant modification in mammalian DNA, as already noted, is methylation of cytosine, followed by adenine and guanine methylation

Chapter 10

Epigenesis, Nuts and Bolts

We have already mentioned stem cells, with emphasis on 'embryonic' stem cells, called also, as previously noted, 'totipotent' cells. That means cells with total potential to develop into any cell in the body, for instance, liver, skin, or neurons and scores of others. This happens through the process of cellular profusion and differentiation. Later cells can produce only a certain group of cells, as with skin or liver for example. These can give rise to cells that can become highly specialized and take the place of cells of their type that die or are lost.

Why is the subject of stem cells important? It is important because of its role in our understanding of the molecular processes of epigenetics. It is not the first step in that understanding; that was probably taken by the scientist who first used the term 'epigenetics,' Conrad Waddington.

Waddington envisioned the necessity for a new stem cell from which to spawn the epigenes that would control the genes themselves. It was these epigenes that could bring about changes that were considerably more responsive to external factors and act much quicker than could the genes. They were, he reasoned, necessarily much more sensitive to the environment than the genes themselves. He envisioned the process in which a necessary new stem cell might operate and illustrated his vision with a sketch.

The sketch showed simply a ball on top pf a hill. At the bottom of the hill are several boughs or valleys. Once the ball rolls down the hill it goes into one of them. Obviously once it has reached the bottom it will stay there unless a way can be found to move it back up, also obviously a much harder job than getting it down in the first place. But if we are to move the ball from one trough or valley into another, it will be even harder than rolling it back uphill to the top.

The meaning is quite clear. The ball at the top of the hill represents the zygote, that single cell that results from the fusion of the sperm and the egg. As the cells grow rapidly in number and differentiate, becoming more specialized, each cell is like a ball that has gone downhill and rested in one of the troughs. An epigenetic cell is attached to the much larger DNA bearing cell, such as the histone, much like a marble attached to a baseball. The DNA bearing cell is never going to turn into another cell type, nor return to the top to give rise to many different cell types. Unless, that is, some means are discovered to make it do so. Or unless nature does it in its own way.

There are many reasons why the scientist, particularly the medical profession would want to do so. The means for putting the ball back at the top, in its position as a stem cell has been discovered. It required lengthy and tedious work to do so. It need not concern us, as obviously no one would want to use that technology to cause a wound or disfigurement to be copied to another person, a 'subject' in the language of our examination of reincarnation. It is interesting nonetheless to know that it has been done and that treatments of various illnesses and injuries with epigenetic processes are in use. There are several medications with epigenetic impact that are already in use or potentially ready, in treating matters such as addiction, anxiety and depression.

But we will leave it at that in favor of examining how it may happen in nature and natural processes that epigenesis can bring about the results that we have seen described in so many cases in preceding chapters.

Different cells have the same DNA blueprint, but they have different molecular, or epigenetic, modifications. Methylation was mentioned as the most important agent of DNA epigenetic modification. It was also the first to be so identified. These differences are transmitted from mother cell to daughter cells during division. Phrased differently when the DNA is copied to form new chromosomes, the same sequences of A, C, G and T are copied. What the epigenetic modifications involves is, whether the gene is expressed, and if so, to what degree. The epigenetic modifications are also passed on, allowing consistency of gene expression from mother to daughter cells. That is why skin cells produce more skin cells, and liver cells more liver cells, etc. (*Epigenetics Revolution*, p. 56).

Now let us look at the role of emotion in bringing about the baffling results we have seen in the preceding chapters.

Chapter 11

The Intrusion of Emotion

Dr. Jim Tucker, in his *Life before Life* (p. 68, 69) has written a thoughtful analysis of how and why, strong emotion can bring about the strange, often grotesque, results we have seen earlier, and even more grotesque ones we shall see later.

He points mainly, but not exclusively, to cases of hypnosis wherein subjects have relived traumatic experiences and, in the process, have developed skin manifestations corresponding with those suggested by the hypnotist. One involved a man having his arms tied behind his back with a rope. During hypnosis he developed deep indentations on his forearms that resembled rope marks. There have also been, writes Tucker, a number of cases that resulted in blisters upon telling subjects that they were being burned, then touching them with a cool finger. Not all of Tucker's cited cases involve hypnosis, and presumably his argument would apply to all such cases whether involving hypnosis or strong emotional impression in the subject's personal experience.

Tucker's point, he describes, as showing that "the mind can produce changes in the body, that given our present state of knowledge, we are unable to explain." By "mind," as previously explained, he does not necessarily mean the brain, but the world of thoughts or consciousness that exists in the brain. It is this last mentioned aspect of Tucker's writing with which I disagree, despite my full agreement with the substance of his view on the broad subject at issue:

It has been shown, Tucker claims, that something survives our death, and that the "something" can enter a fetus. He explains further that if that something can produce physical changes in the body from whence it came, then there is no reason to believe that it could not produce similar, or identical change in the new body that it inhabits. In short, Tucker continues, if that mental aspect carries traumatic memories which produced certain marks or defects on the previous personality, that mental aspect and those memories could produce birthmarks or birth defects on the subject that matched wounds caused on the prior personality (Life Before Life, p 69, 70).

My difference with Dr. Tucker arises chiefly, perhaps only, with the underlying cause or mechanism driving this phenomenon. He believes there is a spiritual dimension to human life, one that transcends the brain (*Life before Life*, pp. 229-231). We can wonder, he writes, if a consciousness or spiritual component brings emotion with it from previous lives, with all of its advantages and challenges.

If Drs. Stevenson and/or Tucker do not in fact see a soul as the required medium, then he, or they, have left the field open for something as yet unexplainable. Dr. Stevenson addresses the subject in an interesting, if rather obtuse, way, toward the end of his 2nd Volume (*R and B*, p. 2083). He says that "we need to imagine a vehicle for memories between lives. For this vehicle I propose the word *psychophore*, which means 'soul bearing'. If such a vehicle exists, it must be made of some substance."

I consider Dr. Stevenson to be a man of remarkably high intelligence and logical in his thinking. But here he seems to want something both ways that seemingly cannot coexist. There must be something of substance, he writes, but that something of substance must encompass a soul. It seems like an awkward fit. In this strange world anything is possible, but our imaginations here are stretched almost to the breaking point.

I believe, on the contrary that the medium may well be the scientific discovery of the process known as epigenesis. It is, as we have seen, that portion of the inheritance process that is open to influence from extraneous sources, from the environment, sources outside of the germ cells, the egg or the sperm.

It is also difficult to grasp, but perhaps not so much so as the dual nature of the psychophore.

He freely admits that he cannot prove the existence of this psychophore, just as I cannot prove, at this time, that either entanglement of atoms, or epigenetics, physical forces whose existence can indeed be proved, are the moving forces at work here. But the history of the world has been episode after episode of guesses, or dogma, about the nature and location of the claimed spiritual force.

The place of God's existence has been chased throughout the universe, and His particular domains or putative powers, one after another, have ultimately been explained by science.

Most civilizations had gods everywhere, holding sway over everything and every aspect of life and nature. Most have now been dethroned in favor of one Almighty, though to each, a different one. The Judeo-Christian God has been located in the heavens, in or above the clouds where no human could dwell- -until they did, and found no evidence of Him there. Then, it was agreed, He (She or It) lived among the stars, seen then only as pinpoints of light. That lasted for most until Galileo, with exponential increases of his vision by his successors, searched that domain and found no spiritual element, only islands of stars we call galaxies and that discovery was only about a century ago.

The heavens, it now appears, no less than earth is red with blood and teeming with its own version of tooth and claw. Big galaxies gobble smaller galaxies, just as big animals gobble the smaller ones. There is birth and destruction in the cosmos, and explosions, called supernovas, beyond our ability to grasp. Peaceful as the starry night may look, the law of the jungle is the law also of the cosmos.

But we hear that He created mankind on Earth to rule, under His dominion, the entirety. His creatures however, often after staying briefly alive by killing and plundering die often painful deaths themselves.

Humans yearn for eternal life. What science has proved thus far is that what some of them have, mostly by pure chance, are memories of another life, usually not their own, memories that last 7 or 8 years in most cases and then are soon forgotten. Many of the subjects are quite bright, though most seem no smarter than the rest of us who have no such memories.

Admitted it must be, that the ultimate disproof will never happen. The Great Beginning is doubtlessly beyond the reach of science. No matter how many ingenious scenarios scientist may advance, there will always be a question, of "what caused that?" Or "What came before that?" Belief in this spiritual force is purely a matter of faith. A non-belief, or belief in materialism is just a lack of such faith.

The ultimate proof is out of reach, something that cannot be proved or disproved. Those who believe in this spiritualism can believe it in any form or locus they imagine. No one description or place takes precedence. The materialists can and do advance a proved description and can draw a picture, in words or diagram, but only so far. For some of us that suffices. But though our minds can perhaps grasp a universe that will have no end, one that had no beginning is out of our reach, and beyond all of our understanding.

PART III

Reincarnations from Hell

Chapter 12

A Strange Spirituality

The last few paragraphs might have been a good point on which to end this tome. What better way than with a discussion of abstract principles about the ultimate nature of human life? A venture into the realm of pure thought. Materialism versus spirituality. But there would be too much left unsaid, or undescribed, of some of the uglier, but highly relevant underbelly of this subject. It is unfortunate that we cannot leave well enough alone. But this paper would not be complete without some reference to the sordid specimens that are part of the core subjects of this superficially high blown discussion. It is part of the story and, if little else, will tell us something about the human condition. I offer it for whatever anyone wishes to make of it.

We start with four cases, all from Burma. Burma (*R and B*, vol. 2 p 1185-6) is where Dr. Stevenson, the dean of researchers in this field, originally spent much of his time. All of the subjects of these four cases are Burmese, born in Burma, with at least one Burmese parent, most with both parents Burmese. None of the parents were of Japanese ancestry, though all four subjects had memories as Japanese soldiers.

The name of Burma was officially changed to Myanmar in 1989 by the ruling military government. The new name means strong and fast. This was well after the gruesome events befalling the prior personalities of these cases. Further, the name Burma is even now in wide use outside of the country. Hence, the name Burma will be used in these summaries. For much of his investigations Stevenson used the services of a Burmese interpreter, who also acted as assistant investigator, and in whom he expressed complete confidence.

Dr. Stevenson speaks in these cases of birth defects, something far more serious than birthmarks. He has reported, among many others, on the following four cases of Burmese subjects. Each of them remembered previous lives and deaths as Japanese soldiers, killed in Burma during the occupation by Japan, during World War II, from 1942 until their defeat in 1945. The previous personalities were, hence, Japanese soldiers but none have been further identified. That is a matter easily understood considering the violently convoluted circumstances of World War II and its aftermath both in Burma and elsewhere.

Stevenson reports on other similar cases from Burma, as well as from India, Sri Lanka, Turkey, Nigeria, and British Columbia. Some of the Burmese remembered lives and deaths that occurred during the British rule in Burma before the war. Others remember from the "insurgency and lawlessness" that followed the Japanese defeat.

According to Dr. Stevenson, relations between the Burmese and their Japanese occupiers were not uniformly, nor generally, very bad, but that there were many individual cases of abuse by the occupiers. The impetus for revenge, after the defeat by the British and attempted withdrawal of Japanese forces was fierce.

Some of the facts described may charitably be described as bestial, but we must always bear in mind that the people involved, their savage cruelty notwithstanding, are humans, like the rest of us, which may be about as bad as anything we might say about anyone.

The following are the four cases. The first two, in chapter 13, are cross-gender cases, and contain certain particular corroboration in addition to that contained in the last two, in chapter 14.

In the course of violent combat, the trappings of civilization can disappear rather quickly. Torture of the enemy often becomes commonplace, as much as revenge for outrages perpetrated, as for useful information. Different cultures seem to have their own peculiar forms of it.

The Burmese seemed to have indulged in their own specialty. A number of the subjects said that their murderers chopped off some of their fingers and toes before killing them. Some said that their fingers had been chopped off when the raised their hands in futile self-defense or in an equally futile plea for their lives. These latter subjects, born decades later, had missing fingers, but not toes. Some had linear grooves on the legs or arms which they attributed to having been tied with ropes before they were killed.

A reminder that, as shown in the chapter on maternal impression, the world-wide incidence of births with the missing of one or more fingers is 1 out of 90,000.

Chapter 13

Two Burmese Cross-Gender Cases

We turn first to the case of, **Ma Win Tar,** a cross gender case. This baby girl was born February 17, 1962, 17 years after the end of the war, in Pyawbwe in upper Burma. Her mother was a Chinese woman, Daw Khin Win. Her father, a Burmese, died before the investigation. The mother's pregnancy with our subject child, including the delivery, were uncomplicated. Ma Win Tar was the 5th of that couple's 6 children.

Our subject infant began walking at about 10 months; speaking at about 1 ½ yrs. Though otherwise healthy, she was born with serious birth defects of the hands. Namely the middle and ring fingers of the right hand were webbed together and only loosely attached, dangling, to the hand. Both those fingers and the other three fingers of that hand had conscription rings, or grooves, on them. Her parents agreed to the advice of the attending doctor that the dangling fingers should be removed and they were amputated when she was three days old. At birth she had not only the grooves on her fingers but on both wrists and the grooves showed the strands of a rope, something that persisted for a little less than 16 years.

Stevenson theorized that the grooves on Ma Win Tar's fingers may have resulted from wounds on the "the Japanese soldier she was remembering," if he put up his hands, either held together to seek mercy, or to ward off a blow from a sword, the weapon would have struck first the fifth, fourth and third fingers, which were the most defective ones on the child, Ma Win Tar, particularly of the right hand.

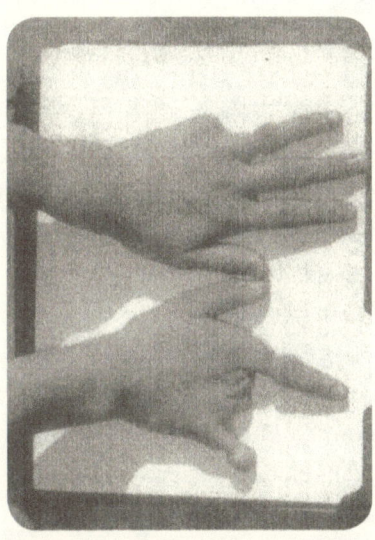

Figure 4: Dorsal and palmar of Ma Win Tar's hands at age 16 in 1978. The three fingers present on the right hand had conscription rings. Swelling can be seen on the right 5th finger. A conscription ring can also be seen on the left forearm. Photo by Dr. Ian Stevenson. Credit: Praeger Publishers.

We might note at this point, that Stevenson and most other investigators often speak of the prior personality as being the life remembered by the subject, and that the subjects themselves refer to them most often as their own 'prior lives.'

Our subject here first spoke about her 'previous life' at about age 3 yrs. She said that she had been a Japanese soldier, captured by Burmese civilians, tied to a tree, and burned alive. Not long after those first sentences on a prior life, she was accidentally scalded with boiling water. At about age 7 years, she developed an intractable mastoiditis the symptoms of which were pain and bleeding. A mastoidectomy was performed. This was not connected to the previous life she described.

Very little does Ma Win Tar, our subject, recall about life in Japan, or her life as a soldier until his capture and execution. By 1977, at age 15 she claimed that before she was scalded she had remembered most of her past life.

In her early years and beyond, Ma Win Tar preferred to dress like a boy and wore trousers until 11 or 12 years of age when her family insisted she dress like a girl. She also liked to keep her hair short like a boy. As a young child she was aggressive and wild, and at times used violence against people who annoyed her. Her mother considered her behavior just a masculine as her dressing like a boy.

Ma Win Tar complained that Burmese food was too spicy and she did not eat spicy food when young but wanted sweet foods. She gradually accepted that she had no choice but to eat Burmese food. She also preferred Chinese or Western dresses to the Burmese longyis.

Her mother said that she was relatively insensitive to pain. She also displayed episodes of cruelty, according to Stevenson something rare among the Burmese. Their treatment of Japanese prisoners, might make that observation somewhat questionable. At times she would slap her playmates in the face, a habit of the occupying Japanese when the Burmese villagers annoyed them. She resisted learning the customs of Burmese Buddhism. When young she sat on the ground with knees forward and buttocks resting on her heels, a Japanese rather than Burmese custom. She also seemed resistant to learning the Burmese language. Even when she did, when angered her voice would show a slight foreign accent.

It was at the age of 15, she was first interviewed by Stevenson. She had not remembered from her life in Japan any names of persons or places, nor how much education the person she remembered had received. From his life as a soldier in Burma she could not remember the name of the enemy the army was fighting, or the name of the place of the last battle before his capture.

But she had remembered a number of things: that he had left his parents and joined his grandfather, a farmer. The grandfather died, and he, at the age of 20 and unmarried, joined the army, and served as a private.

She remembered that he was captured by 10 Burmans and tied to a tree. In vain he pleaded for his life. There were no women among his village captors. The Burmese put faggots around the tree, and set fire to them, burning him alive. She did not know where he had been killed.

There is much circumstantial evidence to support her claim of a 'prior life' as a Japanese national, but none as strong as the birth defects. In this, there is strong corroboration in the accounts of others who remember similar treatment.

*   *   *

We look next at the case of **Ma Win Yee**, another case of cross gender, and one with memories of a Japanese soldier in World War II, executed by Burmese villagers. Like the subject of the previous case, Ma Win Tar, she was the 5th of 6 children. In many other respects this case differs from the previous case.

She was born in the village of Shweda, also in Upper Burma. Her mother was Daw Aye, whose husband, also her first cousin, died before the case was studied by Dr. Stevenson. When she was born, her mother noticed defects of the left hand and of the toes of both her feet. She began to speak at about 2 years of age and almost immediately begin talking about a prior life in Japan, including fighting in the Japanese army and being killed in Shweda. She told Stevenson that she had been married in Japan and had three children there. But she did not remember the sexes of the children, the soldier's occupation before joining the army, where he lived in Japan, or his rank in the army

The area around Shweda was strategically important during the fighting in the spring of 1945 when the British reconquered Burma. One witness said that during the fighting she saw a lot of Japanese dead at the north end of the village. Ma Win Yee's mother said she had not gone near the bodies of Japanese soldiers. Witnesses who heard the subject's statements during early childhood were all deceased. The mother and sister, busy with chores and work, had little time to listen to the child. A retired judge, however, remembered that as a child of 4 or 5, she had told him that she had been a soldier from Tokyo, showed him the spot where she said the soldier had died, and showed him the defects of her fingers and toes. He remembered also that she had said she had been "chopped" on her fingers and toes.

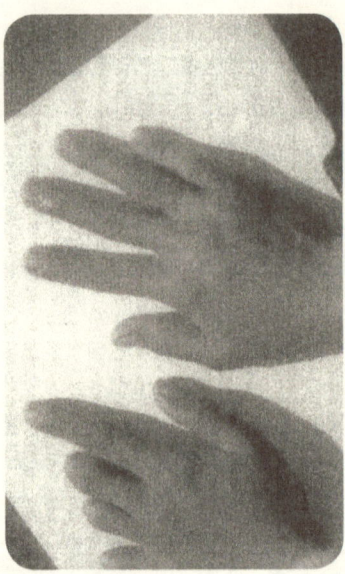

Figure 5: Dorsal and palmer of Ma Win Yee's hands at age 26 in 1980. The middle fourth and fifth fingers of the left hand had no nails. Photo by Dr. Stevenson. Credit: Praeger Publishers

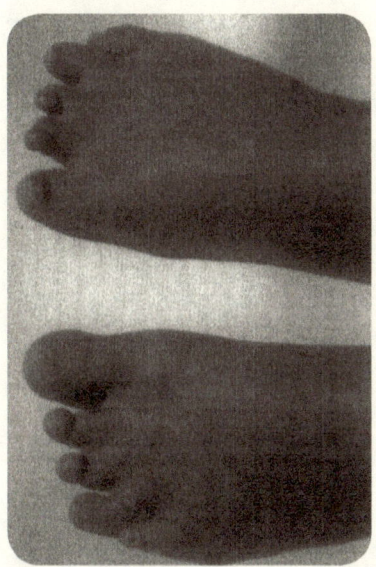

Figure 6: Feet of Ma Win Yee in 1980. All toes were short and most were poorly separated from each other. Only the filth toes and possibly the fourth toe of the left foot had nails. Photo by Dr. Stevenson. Credit: Praeger Publishers

Ma win Yee, unlike others who remember lives of Japanese soldiers, showed nothing of typical Japanese behavior. The only such indicia was her preference for strong tea, whereas Burmese prefer exceptionally weak tea. She recalled thinking before dying that she would never return home to Japan. She now had no nostalgia for Japan, but longed to be reunited with her children. She likewise never asked for boys' clothes, or played boys' games, something markedly different from other girls who remembered lives as Japanese soldiers. When young they showed distinctly masculine traits, in addition to more typical Japanese behavior.

Taking into account the statements of witnesses, including the methods of some Burmese soldiers, Stevenson concluded that what she has recounted is credible, though no Japanese soldier whose life corresponds to her statements has been, or is likely now to be identified. Whatever her defects of memory, the absence of toes and fingers, considered together with evidence of such defects by others, is highly compelling.

Chapter 14

Two Burmese same-gender cases

Another of the cases showing such defects is that of **Maung Aung Htoo.** He was born in January 1962 in the town of Koolon in southern Burma. His father, who died in 1980 was a Chin and lance corporal in the Burmese army. His wife, the mother of Maung Aung Htoo and five other children born after him, was Daw Khin Kyi, a Burmese. Apparently, in that area, relations between the Burmese and the Japanese occupiers were friendly.

When Maung Aung Htoo was born he was noticed immediately to have defects of his right foot and of both hands. Photographs taken by Stevenson in 1980 when the subject was 18 years old show the defects were still there. On the hands only the thumbs were normal. All of the fingers were markedly shortened. The middle and index fingers of the left hand were joined together. On the right hand only the fourth and fifth fingers had nails. Photographs of the feet show that on the left foot the fourth and fifth toes appeared abnormally short. On the right foot only the fifth toe had a nail. All other toes lacked nails and were markedly shortened. The third toe may have been absent, or may have been partly joined with the second toe.

Stevenson also saw clear signs of grooves in the lower part of the left calf and the left lower thigh, indicative of tight bindings.

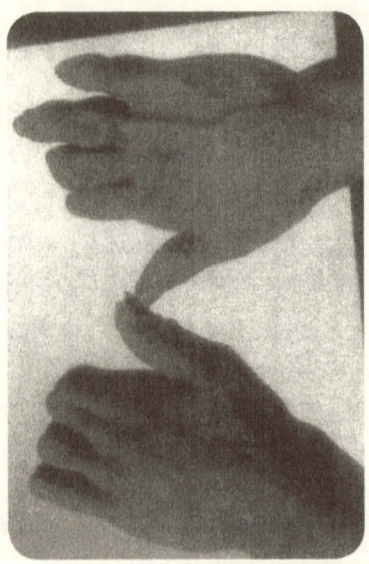

Figure 7: dorsal and palmer of hands of Maung Aung Htoo's hands at almost 19 years, in 1980. Only the fourth and fifth fingers of the right hand had nails. Photo by Dr. Stevenson. Credit: Praeger Publishers

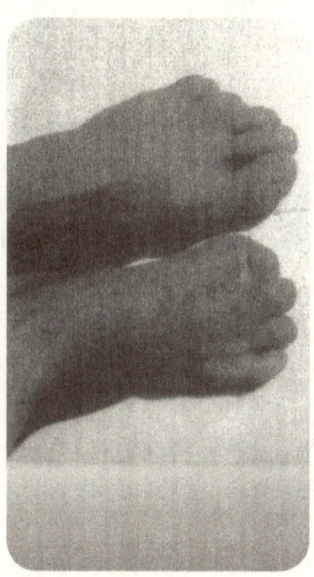

Figure 8: Feet of Maung Aung Htoo in 1980. All toes lacked nails, except for the fifth toe on the right foot. The third toe, presumably, on the left foot, according to Dr. Stevenson was either absent or partially absorbed with the second toe. Fourth and fifth toes of the left foot were abnormally short. All other toes were markedly shortened. Photo by Dr. Stevenson. Credit: Praeger Publishers.

No other family member had a birth defect of any kind. His mother was in good health throughout her pregnancy with Maung Sung Htoo.

His mother said that he began to speak coherently at age two, and was three when he first spoke of his previous life. He said he had killed members of his family, joined the army, and had been a Japanese army officer. When the Japanese army retreated from Burma, the soldiers he commanded mutinied, tortured him by cutting his toes and fingers, and left him tied to a tree where he expired five days later.

In 1980 he told Stevenson that he remembered having insignia with three bars but did not remember his rank. He had been married and had five children, but did not remember their sexes. Before he left Japan he killed his wife, children and two sisters, as he did not want to be distracted thinking of them while fighting. While in Burma he had conscripted Burmese villagers working for him. He "intensely" disliked them, and after making them work he killed them.

Upon their retreat some of the 22 soldiers he commanded mutinied and after the torture, tied him to a tree, and placed a bomb behind it. However the bomb didn't explode, and, he said, he was tied to the tree for five days before he died. Stevenson, a trained physician, wrote that he would have died from hemorrhage, shock, and dehydration and that he, Stevenson, was surprised he could have lived that long. He noted also that the interval he mentioned may be defective, as may other details of his narration, which "are all unverified."

Stevenson noted also that none of the informants he talked to had ever heard of a Japanese officer whose troops had mutinied and tortured him. He acknowledges that it could have happened, but would be unusual behavior for the well-disciplined Japanese. It may be noted here, that if there ever was such a departure from good discipline, this would be a likely case for it. In addition to what we have heard thus far, there are accounts of sadism in his "present" life that seem to corroborate his accounts of his grisly criminal behavior in his "past" life, to which we shall turn shortly.

Stevenson's trusted translator, a Burmese, said that the behavior was more typical of Burmese villagers in those days. Stevenson adds that if the Japanese officer who Maung Aung Htoo recalled was as brutal as he said, he would have been one who the villagers would likely have captured and tortured, the mode of torture being more typical of Burmese than the Japanese.

In the presence of his maternal aunt, he said that when angered he would slap and kick his younger brothers. He would hunt and kill insects, frogs, fish, and birds. The sadism was apparently second nature to him. He would take the legs off a frog one at a time while it was still living. He matter-of-factly described his pleasure in hunting and cruelty to other living things. If the law did not prohibit it, he said he would kill human beings. When Stevenson suggested he might be exaggerating his aunt agreed with his denial that he was. Nor did he, apparently, have any nightmares.

All of this notwithstanding, his aunt also described him as industrious, quick witted, intelligent, and straightforward in his relationships. Nonetheless, due to his behavior he became an outcast in the family. His mother, aunt and grandmother apparently loved him, but did not conceal dismay over his cruelty. From their accounts, his father had been a pious person and kind to animals. When the women in the family tried to inculcate him with the precepts of Buddhism, they were completely unsuccessful.

Of his claim to have been originally Japanese, there is much circumstantial evidence. It showed in his preference for clothing, his taste in food, and his difficulty as a young child in learning to speak Burmese.

His mother said he was hard working and, compared to the other children, insensitive to pain, and would put salt on his wounds. She also said that he held no hatred toward the Japanese who, he said, tortured and killed him in his previous life. He acknowledged the "incompetence" of the man he thought he had been, and that this man might have got what he deserved. He showed no anger when hearing mention of the British or Americans. However, his fierce animosity toward the Burmese people continued. He expressed a strong desire to return to Japan.

There is much supporting evidence of Japanese nationality, incontrovertible evidence of torture, though not necessarily by Japanese soldiers as he claims. Other cases summarized here indicate, as does the opinion of Stevenson and his Burmese assistant, that it might well have been the Burmese villagers. We will look at one more case of the same kind of torture, for any bearing it may have on this issue.

*   *   *

Maung Hla Hsaung was born in the village of Nan-U Lwin in upper Burma in October 1960, the fifth of seven children of U Pu and his wife Daw Hla Nyein. U Pu died in 1970. There have been a number of cases including 'announcing dreams,' something almost self-explanatory. I have omitted mentioning them previously as not necessary to the essence of the case. I mention the one in this case as I feel it may have a more than usual relevance.

A month or two before the mother became pregnant with the above subject she dreamed of a man she did not recognize. He had defects of his hands and feet as though they had been chopped off. She did not recognize his nationality. Upon awakening she told her husband that their next child would be defective. Later, in 1984, she told Stevenson that she connected the dream with a pregnancy and thought that the man wanted to be her child as the dream was more vivid than other dreams and that she had no dream with previous pregnancies.

When our subject, Maung Hla Hsaung was born the midwife immediately noticed the defects of his hands and one foot. She seemed determined to keep this news from the mother, wrapping the infant in a blanket, telling the mother not to unwrap it even when nursing it, and warning the family members that the mother might be shocked by the sight of the defects. Hence it was some time before the mother saw her child unwrapped.

Maung Hla Hsaung started speaking at about 2 and was not yet 3 when he started speaking, mostly to his father, about a previous life in Japan. Stevenson and his translator did not start investigating this case until 1984, by which time the father was deceased. Other family members remembered very little, but did recall that he spoke a strange language for some time, then learned Burmese normally. He claimed to have been a Japanese soldier and that the defects of his fingers and toes were the result of torture after capture. He claimed that the torture had been by the English.

When Stevenson began his investigation in February 1984, Maung Hla was about 23 years old. As was Stevenson's custom he photographed the defects and described them in detail. The fifth finger of the right hand was very much shortened and had no nail. The other fingers of that hand were normal except for grooves at the bases of the middle and fourth fingers, indicating tight binding. On the left hand only the thumb was normal. The other fingers were shortened and had no nails. At birth the fourth and fifth fingers of the left hand were joined and dangling as if by strings and were amputated when he was about 18 months old.

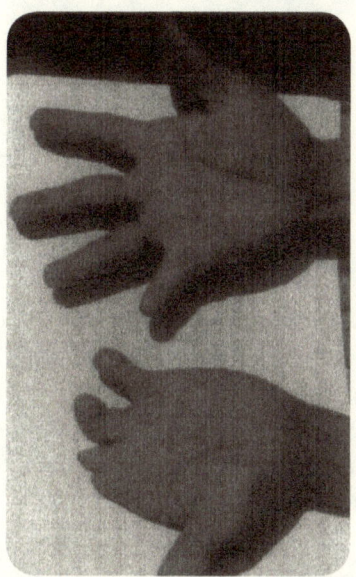

Dorsal and palmer of hands of Maung Hla Hsaung at age 24 in 1984. The fifth finger of right hand was much shortened and had no nail. On the left hand, all fingers, except for the thumb were shortened and had no nails. Photo by Dr. Stevenson. Credit: Praeger Publishers.

The left foot was normal, though the toes of the right foot, except for the fourth toe, were markedly shorter than those of the left foot, and had no nails. The first three toes of the right foot were joined and the right big toe was almost absent. In addition there was significant evidence of the legs having been tightly tied by ropes.

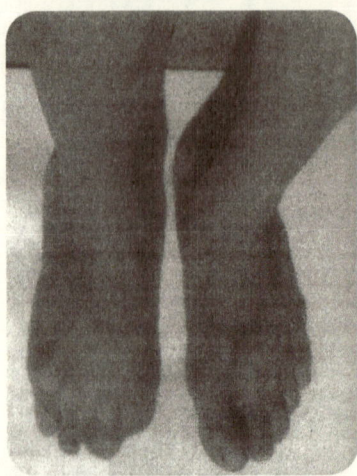

Figure 10: Maung Hla Hsaung's feet at 24 years of age in 1984. The toes on the left foot were normal, as was the fourth toe on the right foot. The remaining toes on the right foot were markedly shorter than the corresponding ones on the left. Also they had no nails and the first three were joined. The right great toe was almost absent.

There was much in his behavior that tends to corroborate his claim of a former Japanese life. When young he would often talk about Japan. When he was given money, or when he asked for money he often said that he was going to Japan. As a young child he sometimes spoke a language that family and family friends did not recognize. He, though not his brothers, used also to play at being a soldier, at which time he wore a Japanese army cap he had found. The clothes he wore and the food he liked showed a preference for the Japanese tastes rather than the usual Burmese customs. He had a preference for sweet foods, and liked pork cooked with sugar. He was comparatively insensitive to pain. He was described by his mother as hard working and independent compared with Burmese. He, himself, said that he had a tendency toward cruelty to animals, though not to humans. In his interview with Dr. Stevenson in 1984 he said that he still longed to go to Japan but that it was not very strong.

Interestingly, his mother said that during her pregnancy with Maung Hla Hsaung she had a craving for pork cooked with sugar.

It is interesting also that Dr. Stevenson seems to have considerable difficulty in accepting his claim that it was the English who inflicted the torture. Stevenson acknowledges that there are cases of cruelty, sometimes of torture by the allies, but he states that the instances are rare and that as far as he knew they never used the Burmese form of torture, namely the chopping of fingers and toes. Stevenson then indulges in some speculation as to how it may have come about that the subject in this case had come to believe that the English were his torturers.

There appears no point in examining these scenarios, as for our purposes it would not matter who did the deeds. It is clear that the injuries were inflicted by someone, and that, as so often, the innocent somewhere suffer for the sins of others, and for still others who may not themselves been the sinners, but wore the uniform of those who were. 'Reincarnation,' no matter the name, or explanation for it, may not always be such a wonderfully spiritual thing.

These four cases are offered as corroboration for what might be doubtful if reported by only one. The next case is reported and summarized here for its uniqueness.

Chapter 15

A Unique Case

U Tint Aug was born in Upper Burma in August 1930. He was the first of 13 children born to U Thu Daw, who died in 1955, and his wife, Daw Hla Po. Stevenson writes that after U Tint Aug's birth the babies came as fast as babies can and he was therefore raised by one of his aunts, the mother's sister in law, Daw Thin. Before U Tint Aug's birth, Daw Thin had reportedly dreamed about the rebirth in their family of U Thu Daw's younger brother, hence our subject's uncle, named Bo San Pe.

Stevenson describes Bo San Pe as a notorious 'dacoit' for about 2 years before he was executed by hanging in April 1929. 'Dacoit' is a term often used in parts of India and in Burma, and refers to a class of criminals in those states who rob and murder in roving gangs. He was deemed to be the prior personality for rebirth in the person of our subject, U Tint Aug.

U Tint Aug had been born with an unusually small mandible, namely the lower jaw bone, one with which he could not fully open. The medical term is 'micrognathia.' The opening of the mouth was correspondingly small, a condition called 'microstomia.' This convinced the family that he was the reincarnation of Bo San Pe. The family also believed they saw a birth mark on the right side of his neck a survival of the hangman's rope. Because the hangman bungled the first hanging, it was necessary that he be hanged a second time, which went as planned.

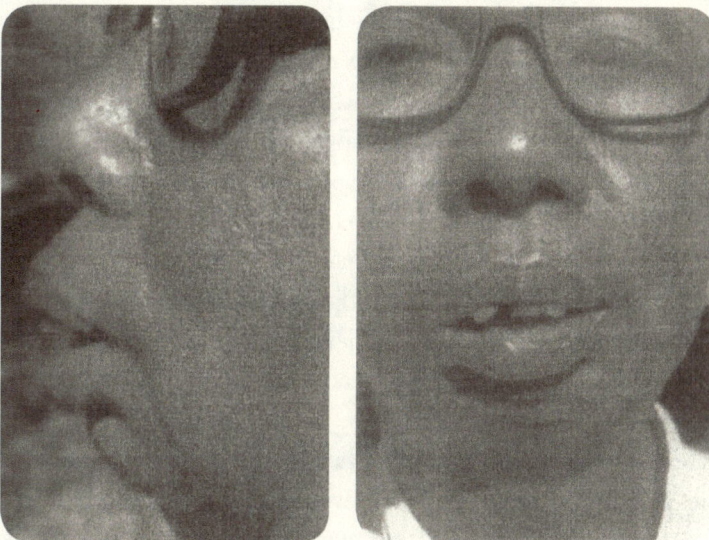

Figure 11: Profile and full frontal photos of U Tint Aug's face at age 53 in 1984. Photos show the pronounced undevelopement of the frontal part of the mandible causing the extreme degree of receding chin. The full frontal photo shows the full extent of the opening of the mouth. The subject is shown here trying to open his mouth as widely as he could. Photo by Dr. Stevenson. Credit: Praeger Publisher.

U Tint Aug began talking coherently at about age 3 and some of his talk concerned the life of Bo San Pe. Stevenson and his assistants questioned and otherwise investigated all know sources for verification, and Stevenson offers an unusual admonition. The sources, he says, should be treated with reserve. Some of what he narrates, he continues, should not be recorded as accurate events but "as representative of what was told and believe about Bo San Pe in his time and afterward."

Stevenson then recites a rather lengthy narrative about Bo San Pe's life of victimization by others, unfaithfulness of his wife, and his revenge, first upon his tormentors and ultimately to serial murders and robbery. His dacoit was of his own making, and he was obliged to keep the members faithful. He earned some support among the populace for his Robinhood-like tendency to sometimes rob from the rich to give to the poor. He was bold and daring, but succumbed to capture when, weak through sickness, he was unable to resist.

He was tried and sentenced to be hanged until dead. In his interview with Stevenson in 1984, U Tint Aug claimed to remember his hanging as Bo San Pe. Stevenson asked him whether his chin had been damaged during the first hanging. He was supposed to have dropped through a trapdoor when it opened so that his neck would break. But the rope had slipped at the beam where it had been tied. So the prisoner would have fallen through the trapdoor to the ground. He said his jaw was not damaged. He had however told one of the investigators that during the first hanging he did have a tight sensation in the jaw. From this Stevenson concluded that he might have still been experiencing this tight jaw during the second hanging from which he died instantly. If so he continued, "the sensation would have been among Bo San Pe's last thoughts before dying."

That fact, in the overall picture Stevenson was trying to recreate, would have meant that his memories, that may have entered another fetus such as his nephew U Tint Aug, would have impacted that subject's body as explained by Stevenson and even more directly, and explicitly by Jim Tucker.

Stevenson said that when he examined U Tint Aug he saw no evidence on the neck that corresponded to a rope mark. But in 1984 this subject was over 53 years old. His former school teacher however said such a mark was clearly visible when he had seen it many years earlier.

He was educated in a monastery school and qualified as a teacher. He married and had children. Nonetheless, it should not be surprising that the disfigurement of his face had a profound negative impact. The furthest he could open his mouth was about 5 millimeters. He was obliged to eat semiliquid food and he tried to avoid eating in public or before strangers. He had many depressive episodes when young and was sometimes reduced to crying spells.

Though he was said to be generally quiet, as a young child he was also said to be quickly angered, and sometimes enraged. U Tint Aug himself said that as a student in school he carried a knife and a razor blade. When schoolmates teased him about his appearance, he would cut them with the razor blade. One of his former classmates told the investigators that when he teased U Tint Aug about his claim that he was Bo San Pe, he did exactly that. He cut him with a razor blade.

Not all reincarnations come from heaven. This one seems to have come from hell.

Chapter 16

So Where Does All That Leave Us?

For those souls who believe in souls, it is all proof of the existence of the soul. But not everyone does. For the rest of us is there anything in all of this to be learned? Probably so, but what it is, I cannot yet guess, but will continue to anticipate the results of further scientific research.

So far, it seems like there are two elements that we need before making even a tentative judgment about what is going on with these strange happenings. We limit 'strange happenings' here to just that subsection of them that has been treated in this paper: The physical similarities between injuries to the prior personality and birth marks or birth defects on the subject.

According to the insight of Drs. Ian Stevenson and Jim Tucker, there are two primary elements. The first requirement is the emotional fixation of his or her injury by the prior personality. Apparently the horror of the realization that death is upon him or her even for a matter of seconds, can sometimes be sufficient. Any physical or psychological trauma resulting in prolonged agony is also sufficient. The second requirement is the attachment to a subject, usually a fetus or newborn, of the memories and associated emotions of the prior personality. As to some subjects that mental element attaches certainly by chance, pure happenstance.

Some young children with memories of other lives, claim that they chose the parents to whom they were born. About that, we cannot be so sure. The credibility of their memories of another life can be investigated and confirmed. That is what makes it believable to so many. There is no means of verifying claims of choice by the entity from the realm 'in between.' For this reason I have chosen not to deal with it in this or prior writing, though in a few cases I have noted the claim. There is so much conjecture and pure guesswork in this area however, I feel that my input, if not the best, would probably not be the worst.

I had noted the obvious, namely that often the subject and prior personality did not even know each other. It is obvious also that many other times they did, and many times were closely related. That much is fact. Why that is, though no one can really know, should not be hard to accept on one piece of guesswork or another.

As I stated and quoted the works of others in my previous paper, *Something Survives*, we breathe the air, at least a few molecules of it, of almost everyone who ever lived, whether humans, animals, or plants. We digest atoms of carbon and other elements that have been in ancient forests, and that may later wind up in our food, or in a painting on someone's canvas. Listen to this outpouring from Lawrence Krauss in his excellent book entitled *Atom* (p.285-6):

> *Can each of the atoms in the air I breathe really have gone through hell and back, braved the bitter cold of space, the brutal heat of stars, have crashed into Earth, have dredged down below the continents and ocean floor merely to rise again? Have these atoms been a part of countless lives, and seen countless deaths? . . . I am only a temporary abode, and my life is an inconsequential moment in their vast eternity.*

And these are atoms that are not necessarily, and most are probably are not, 'entangled.' The atoms that comprise our organs of memory, most probably the hippocampus and the frontal of the brain, may well be entangled and can and may communicate with each other, some of them probably for untold eons after the Earth no longer exists, as their eternity is indeed vast. They were formed in the stars billions of years ago, and the projected existence of some percentage of them is billions of times longer than the projected life of the known universe.

The trillions of atoms in each of our bodies has all of the attributes and characteristics usually ascribed to the soul. In such a realization, the existence of souls is not only elusive, but also unnecessary. Atoms survive the death of the bodies they inhabit. They live (virtually) forever. Many communicate with each other.

The proven facts outlined in these pages, even after all the conjecture and guesswork are dismissed or ignored, already tell us, as do so many other scientific studies, that what is 'out there' to be discovered dwarfs what we already think we know.

Bibliography

Aczel, A. *entanglement*. New York: Penguin, 2001.

Bernstein, J. *Quantum leaps*. Cambridge, MA: Belknap Press, 2009.

Carey, N. *The Epigenetic Revolution*. New York: Columbia UP, 2012.

Gilder, L. *The Age of Entanglement*. New York: Vintage Books, 2009.

Haraldsson, E. "Children Claiming Past-Life Memories," in Journal of Scientrjic E,rploratron, Vol.5, 1991, 235-243.

Holt, J. *Why Does the World Exist?* New York: Liveright Publishing, 2012.

Jablonka, E. and Lamb, M.J. *Epigenetic Inheritance and Evolution*. New York: Oxford UP, 1995.

Jones, W. *Habit and Heritage*. London: Kegan Paul, Trench, Trubner, 1943.

Koestler, A. *The Ghost in the Machine*. London: Hutchinson, 1976.

Krauss, L.M. *Atom*. Little Brown, 2001

Mitchell, E. *The Way of the Explorer*. Franklin Lakes, NJ: New Page Books, 2008.

Moore, D.S. *The Developing Genome*. New York: Oxford UP, 2015.

Pasricha, S. *Claims of Reincarnation*. New Delhi: Harman Publishing House, 1990.

Rane, W. *Soul Hunters*. Berkeley, CA: University of California Press, 2007.

Rosenblum, B. and Kuttner, F. *Quantum Enigma*. New York: Oxford UP, 2011.

Shroder, T. *Old Souls*., New York: Simon & Schuster, 1999. Payne. New York: Dover Publications, 1966.

Skrbina, D. *Panpsychism in the West*. Cambridge, MA, 2007.

Spiro, M. E. *Burmese Super-Naturalism*, Expanded Ed. New Brunswick, NJ: Transaction Publishers, 1996.

Stevenson, I. *Children Who Remember Previous Lives*. Jefferson, NC: McFarland and Company, 2001.

Stevenson, I. *European Cases of the Reincarnation Type*. Jefferson, NC: McFarland and Company, 2003.

Stevenson, I. *Reincarnation and Biology*, two volumes. Westport, CT: Praeger, 1997.

Stevenson, I. *Twenty Cases Suggestive of Reincarnation*, 2nd Ed, revised and enlarged. Charlottesville: University of Virginia Press, 1974

Stevenson, I. *Where Reincarnation and Biology Intersect*. Westport CT: Praeger, 1997.

Talbot, M. *The Holographic Universe*. New York: Harper Collins, 1991.

Tucker, J. B. *Life before Life*. New York: St. Martin's Press, 2005.

Turkheimer, E. and Gottesman, I. "Individual differences and the canalization of human behavior", in *Developmental Psychology*, Vol. 27(1), Jan 1991.

Waddington, C.H. "Evolutionary Adaptation," in *Learning, Development and Culture*. Ed. H.C. Plotkin. Chichester, NY: John Wiley & Sons, 1982

Waddington. C. H. *The Evolution of an Evolutionist*. Ithaca: Cornell UP, 1975.

Zohar, D. *The Quantum Self*. New York: William Morrow, 1990.

www.ingramcontent.com/pod-product-compliance
Lightning Source LLC
Chambersburg PA
CBHW030906180526
45163CB00004B/1720